T0076760

A LITTLE BOOK ABOUT

# THE BIG BANG

A LITTLE BOOK ABOUT

# THE

# BIG

# BANG

**Tony Rothman**

THE BELKNAP PRESS OF HARVARD UNIVERSITY PRESS
Cambridge, Massachusetts · London, England
2022

*Library of Congress Cataloging-in-Publication Data*
Names: Rothman, Tony, author.
Title: A Little Book about the Big Bang / Tony Rothman.
Description: Cambridge, Massachusetts: The Belknap
Press of Harvard University Press, 2022. |
Includes bibliographical references and index.
Identifiers: LCCN 2021032543 |
ISBN 9780674251847 (cloth)
Subjects: LCSH: Big bang theory. | Cosmology. |
Quantum theory. | Dark energy (Astronomy).
Classification: LCC QB991.B54 R68 2022|
DDC QB991.B54—dc23
LC record available at https://lccn.loc.gov/2021032543

*To my professors and colleagues,*
*who taught me more than they know*

# CONTENTS

## ALSO BY TONY ROTHMAN

A LITTLE BOOK ABOUT
# THE BIG BANG

# WHY IS THERE SOMETHING RATHER THAN NOTHING?

This is a little book on the biggest subject conceivable—the big bang. It is not a book about a television show. It is a book about cosmology. Cosmology, as cosmologists think of it, is the study of the structure and evolution of the universe as a whole. Over the past century, it has increasingly come to mean the study of the early universe: investigation of the origin of galaxies, analysis of the lightest chemical elements, observation of the heat radiation pervading all space, and exploration of exotic phenomena we can't directly see— dark matter and dark energy. Generally, cosmologists concern themselves with our universe in the first eons, years, and even fractions of a second after its birth. Cosmology is precisely the theory of the universe's origin: the big bang.

Cosmology is occasionally called the place where physics and philosophy meet. That is to an extent true, and to an extent unavoidable. When we get down to it, all science is the asking of questions and the pursuit of answers to those questions. If we pursue the questions far enough, we inevitably run out of answers. Cosmology is uniquely prone to this difficulty. When a conversation arises about the big bang, the first question any non-cosmologist (which is most people) asks is "What came before the big bang?" This is a natural and legitimate question, but it presently has no answer and that state of affairs is likely to persist past the shelf life of this author.

Nevertheless, my plan is to pose the questions asked by laypersons, as well as others, and attempt to answer them in the simplest manner I can. Since this is a book meant primarily for people who are curious about science but lack scientific and mathematical backgrounds, my colleagues will find it equally lacking in rigor and completeness, but my aim is not to cover as much territory as possible; rather it is to uncover a little territory if possible.

To that end, I have tried to keep technical jargon to a minimum, and although there will be enough numbers to satisfy anyone, no equation in the text is more complicated than one for a straight line; anything else, I've relegated to the few footnotes. I also assume that readers can understand basic graphs and are willing to follow some fairly detailed arguments. On the other hand, I agree with one of the countless aphorisms Einstein never uttered, "You should make things as simple as possible, but not too simple." Over the years, I have become convinced that there really is a level below which certain things cannot be simplified; in cosmology this is largely because of its inherently mathematical nature. If I cannot explain the mathematics in terms of a comprehensible physical concept, I won't try.

Despite the lack of anything resembling real math in this book, one of its aims is to convince you that modern cosmology is an extraordinary edifice built on rock-solid foundations and that you should become a believer. To that end, each chapter generally builds on the previous. You should start the book at the beginning. If your only interest is the bottom line, you will grow impatient.

As I've said, cosmology does raise profound questions. In exploring the conceptual under-pinnings of the modern big bang theory, my hope is not to shy away from such questions. As a mentor once advised, "If you ask a stupid question you may feel stupid. If you don't ask a stupid question, you remain stupid."

Inevitably, as the book progresses there will be more questions than answers. After all, in pondering the imponderable it is a short leap from "What came before the big bang?" to the ultimate conundrum: "Why is there something rather than nothing?" Given that people have been asking this question one way or another for millennia without consensus, it is not reasonable to expect to find the answer here. Indeed, if you put that question to any honest cosmologist, the only reply you will get is "I don't know." An easier question is, "Do those equations on the white board of the TV show mean anything?" The answer is yes. Personal experience suggests that cosmologists are underequipped to answer any questions regarding cosmetics.

*

Because this book is intended for general readers, I will make use of analogies rather than equations. A danger lurks here because sooner or later every analogy breaks down. Analogies, like theories, are models of reality, not reality itself. In the case of the big bang, cosmologists usually resort to balloons to explain certain properties of the expanding universe, but the real universe is not a balloon and the analogy is imperfect. When considering analogies, it is crucial to locate the differences between the analogy and the reality.

I have already used the word *theory* several times. Let me emphasize that when a scientist uses this term, it carries a different meaning than in daily life. The radio often informs listeners that a prosecutor has a certain theory about a crime, while the defense attorney has a theory that the prosecutor is crazy. Usually, these are conjectures made entirely without evidence and the situation changes too frequently to make any sense of it.

By contrast, a physical theory is a highly interconnected web of ideas and predictions underpinned by mathematics and firmly supported by experimental and observational evidence. When cosmologists speak of the big bang theory, they are referring to just such a web of predictions and observations. The elements of the big bang theory have by now been under scrutiny for an entire century, and so many precision observations support the overall picture that some cosmologists feel that their discipline already resembles engineering more than it does basic research. Believe in modern cosmology.

\*

Yet a fundamental difference between cosmology and most other sciences remains: There exists a single observable universe. The essence of most sciences is experimentation and replication. A drug manufacturer tests a vaccine by running clinical trials on many subjects. If the results cannot be reproduced by scientists worldwide, the vaccine is not regarded as reliable. Cosmologists, at least at present, are denied the opportunity to run experiments on multiple universes and thus they cannot say with complete certainty how the universe would look had things started off differently than they did.

Nevertheless, although cosmologists can't say everything, they can say far more than nothing. Having a single universe at our disposal only makes it difficult when considering the universe as a whole, when addressing ultimate questions. Short of that, cosmologists draw on data and observations collected by their close cousins, the astronomers. Astronomers have traditionally investigated the behavior of planets, stars, and galaxies through earthbound telescopes or telescopes in near-earth orbit. Yes, astronomers are landlubbers, or might as well be; no spacecraft or telescope has yet traveled anywhere near the distance to the next star, yet alone another galaxy, which means it is impossible to perform experiments on astronomical objects. For good reason astronomy is termed an observational science.

The basic assumption underlying all astronomy, however, is that the fundamental laws of physics are the same throughout the universe. Astrophysicists, also close cousins to cosmologists and astronomers, have applied these laws to decode the behavior of stars and galaxies. Since it is impractical to send a space probe to the distant reaches of the universe, at least within the lifespan of a civilization, we have instead relied on light and other messengers to bring information from the far universe to us. It is, in fact, one of the great triumphs of modern science that we have been able to learn so much about the cosmos without going anywhere, by making this assumption that the laws of nature as we know them apply everywhere. To what extent the known laws of physics apply to the universe as a whole remains an open question.

Cosmologists attempt to reconstruct the evolution of the universe using the same approach as astronomers and astrophysicists: with pen and paper or computer, we apply established physics in a mathematically consistent way to model the system we are studying and check whether the results agree with observation. The system may be a cluster of galaxies or the whole universe. If the predictions of our model agree with the observations, we go out for a beer. If the predictions don't agree, we search for mathematical mistakes. If we find none, we search for conceptual errors. If, finally, no one's model agrees with the observations, we add new phenomena. If the new phenomena improve the results, we ask our observational colleagues to begin a search.

One thing any scientist should hesitate to do is add exotic phenomena to the current model before having exhausted more pedestrian explanations. In thinking about the earliest instants after the big bang, hmm. . . .

✳

At this moment you may be wondering exactly where astronomy and astrophysics leave off and cosmology begins. There is no precise boundary, and typically a scientist working in one of these areas knows a fair bit about the others. The difference is mainly one of *scale*. As mentioned, astronomy and astrophysics are traditionally concerned with the behavior of stars, planets, and galaxies, more recently with entire clusters of galaxies and even the superclusters—clusters of clusters of galaxies. A cosmologist takes the biggest picture imaginable, which begins somewhere around the size of a supercluster and asks how all this came to resemble the universe we observe. Although the physics governing the behavior of galaxies is the same as for stars, this book will not be concerned with those, or with planets. It will barely touch on black holes, as fascinating as they are. From a cosmological perspective, these objects are so small as to be insignificant.

Cosmologists find it extremely helpful to keep in mind the various astronomical scales. Throughout the book I will use the standard astronomical practice of stating distances in terms of the time it takes light to travel those distances. You may know that it takes light about eight minutes to travel from the sun to the earth. Call it ten. We can thus say that earth lies at a distance of about ten light-minutes from the sun. Similarly, a light-year is simply the distance light travels in one year. Astronomers never convert light-years to miles or kilometers, and you shouldn't, either. Rather, you should just develop a feel for the different scales found in the universe:

Four light-years is the distance to the nearest star beyond the sun.

The diameter of our Milky Way galaxy is roughly 100,000 light-years.

The distance across a cluster of galaxies is millions of light-years.

The size of a supercluster of galaxies is hundreds of millions of light-years.

The size of the observable universe is about fourteen billion light-years.

✳

That is the scale of cosmology, the scale with which this book is concerned.

**Can you give me advice on eyeshadow and mascara? No.**

# GRAVITY, PUMPKINS, AND COSMOLOGY

COSMOLOGY IS the study of how gravity determines the evolution of the entire universe, so to understand cosmology requires understanding gravity.

Gravity is by far the weakest of the known natural forces. To a physicist, a force is nothing more than a push or a pull exerted on an object—no "dark side" enters the picture—and one of the main reasons that physicists call their field the most fundamental of all sciences is that, over the centuries, they have learned that only four fundamental forces exist in nature. One of these, termed the *strong nuclear force,* is easily the strongest natural force and holds the nuclei of atoms together. Any atomic nucleus consists of neutrons and protons, and the electrical repulsion among the positively charged protons would cause the nucleus to fly apart were it not for the strong force binding it together. The energy associated with the strong force is what is released in atomic explosions. The strong force, however, operates only within the atomic nucleus, which is extremely small, as cosmology goes.

The second fundamental force is the *weak nuclear force*. Billions of times weaker than the strong force, it governs certain forms of radioactive decay. Tritium, the extra-heavy version of hydrogen, is radioactive and decays into a form of helium; its rate of decay is determined by the weak force. But like the strong force, the weak force operates only within the atomic nucleus, which is insignificant on the scale of cosmology.

In daily life the most important forces are the electric and magnetic forces, which are actually two aspects of a single *electromagnetic force*. This force is responsible for all of chemistry and operates in any device requiring electrical currents, from toasters to smartphones to everything we take for granted today. The electromagnetic force is the basis of modern civilization. But to produce electric or magnetic forces requires electric charges. Because astronomical bodies, such as planets, are electrically uncharged they exert no electrical or magnetic forces on each other.

All objects do gravitationally attract one another. Gravity, though, is almost unimaginably weak—that the gravitational tug of the entire earth cannot budge a refrigerator magnet is a hint of how weak it is compared to the electromagnetic force. The way physicists tend to state it is that the gravitational attraction between two hydrogen nuclei, protons, is about thirty-six orders of magnitude smaller than the electrical repulsion between them. In designing consumer electronics, engineers pay no attention to gravity.

Yet, because nuclear forces operate only inside atomic nuclei and because astronomical bodies are electrically neutral, it is left to the weakest force in nature to determine the fate of the universe.

＊

Our modern theory of gravitation is Albert Einstein's general theory of relativity, which is often called the most beautiful scientific theory. This is true.

On a superficial level, we might regard general relativity as merely a refinement of Newton's theory of gravity, devised by Isaac Newton nearly four hundred years ago. It consists of a single immortal equation that shows how the gravitational force between two objects depends on their masses and the distance separating them. We don't even need to write the equation down to understand its message: knowing just the masses of the objects and their separation allows us to determine exactly the gravitational force they exert on one another.*

Above I said a force in physics is simply a push or a pull. More precisely, a force causes an object to change its velocity—in other words, to accelerate. If a piano is speeding up or slowing down, a force is acting on it. If the piano is moving at a constant velocity, no force is acting on it.

---

* For reference, Newton's law gives the gravitational force $F$ between two masses, $m_1$ and $m_2$ as $F = Gm_1m_2 / r^2$, where $r$ is the distance between them and $G$ is the *gravitational constant,* a number that must be measured in the laboratory and that determines the strength of the force.

According to Newton, if we know the forces on an object, we know its acceleration, and can then completely predict its future behavior. Thus, if we knew the masses and present separations of all the stars in the universe, we would know everything there is to know about the universe's future—and its past, as well. For this reason, the Newtonian universe is often compared to clockwork. For the most part, it is.

✳

Newton's theory of gravity works so well in ordinary circumstances that for two centuries astronomers believed it completely explained the motions of the solar system. In the mid-nineteenth century the first hints appeared that this might not be so. Like all the planets, Mercury travels around the sun in an elliptical orbit. If Mercury and the sun constituted the entire solar system, the point of Mercury's closest approach to the sun, called its *perihelion,* would always remain at a fixed point in space. Astronomers observed instead that the perihelion was gradually shifting its position over time. Calculations indicated that the gravitational tug from the other planets in the solar system could account for most of this shift, but a tiny amount was stubbornly left over. Many theories were proposed to explain the anomaly, but the ghost in the machine remained a mystery for over half a century.

When Einstein began work on general relativity in the early twentieth century, apart from Mercury's perihelion shift there was no observational evidence that Newtonian gravity might be inadequate. There was, however, James Clerk Maxwell's theory of the electromagnetic field.

You should first realize that Newton's theory is one of *particles* and *forces*. Two pumpkins sit in a pumpkin patch. We can think of them as two particles exerting a gravitational force on each other across the patch. Likewise, we can idealize the earth and moon as particles exerting a gravitational attraction on each other across space. In neither case does Newton's theory explain how the force travels from one particle to the other. For this reason, Newtonian gravitation is often called an *action at a distance* theory, *action* being the word for force in Newton's day.

Equally important is that the gravitational force between the two objects is evidently transmitted *instantaneously;* if the sun disappeared, nothing would be left for the planets to orbit and they would fly off into space with no delay whatsoever.

\*

Instead of a pumpkin patch, imagine that the pumpkins are floating in a pond. We immediately feel the picture has changed. The water in the pond is composed of an enormous number of molecules, but they are so tiny we forget about them and instead think of the water as having a certain density and pressure at each point. Density and pressure are "bulk" quantities, making no reference to individual particles. This is a signature characteristic of a *field*. The air in a room can be regarded as a field. So can the elastic surface of a trampoline. A swarm of bees in many respects resembles a field.

The field picture provides a natural mechanism for transmitting forces. If the pumpkins are bobbed up and down, they create small disturbances that propagate across the pond as water waves. These waves are local disturbances traveling through the water field at finite velocities. By contrast, in Newtonian gravity, one needs to imagine forces that are somehow transmitted across great voids, infinitely fast.

"Objection!" you cry, politely: the gravitational attraction between the earth and the moon does not involve waves. True. All analogies break down. When thinking about the permanent gravitational attraction between bodies, whether we imagine forces or fields doesn't much matter. Nevertheless, fields exist; if you have ever sprinkled iron filings onto a piece of paper above a magnet, you have perceived the shape of its magnetic field fairly directly. On the whole, the field picture is so powerful that essentially all modern theories of fundamental physics are field theories. Without the field concept it becomes virtually impossible to describe electromagnetic and gravitational waves.

To be sure, when Maxwell considered the laws governing electric and magnetic fields, he was able to show that these fields could propagate through the vacuum of space in the form of an electromagnetic wave traveling at $3 \times 10^8$ meters per second.* His discovery, published in

---

* Scientific notation is indispensable in physics and astronomy. To clarify for anyone unfamiliar with it, the exponent indicates the number of powers of ten, or how many zeros follow the one. Thus, 10 can be written as $10^1$, 100 as $10^2$, and 1,000 as $10^3$. $3 \times 10^8$ is 300,000,000, which shows why we use scientific notation.

1865, astounded Maxwell, because that number was almost the exact speed of light, which by then had already been accurately measured. The conclusion was "scarcely avoidable," he wrote, that *light itself* must be an electromagnetic wave traveling not infinitely fast but at the finite velocity of $3 \times 10^8$ meters per second. Maxwell's prediction, the greatest theoretical triumph of nineteenth-century physics, was confirmed several decades later by the discovery of radio waves.

At the opening of the twentieth century, a number of physicists attempted to create field theories of gravity based on Maxwell's electromagnetism. They all failed, because gravity doesn't behave exactly like electromagnetism. Einstein was the first to understand the difference and the first to get gravity right. To appreciate how his theory, which he called general relativity, describes the gravitational field, we must first get a feel for the theory he had developed earlier that serves as the point of departure for general relativity: the special theory of relativity.

### What is relative and what isn't?

# A SPECIAL THEORY

FROM THE 1820s onward, natural philosophers understood that electricity and magnetism are intimately related. Electrical currents produce magnetic fields and vice versa. With his theory of electromagnetism, Maxwell showed precisely how this took place. In creating his special theory of relativity, Einstein showed that electricity and magnetism were not only related but were two aspects of the same phenomenon. In doing so, he discovered that Newtonian physics must be modified.

But Einstein would never have agreed with the famous adage "everything's relative." At bottom, virtually all physics concerns motion and the essential question asked by relativity is: What changes when something's state of motion changes, and what stays the same? Some things change while others remain the same, and the theory of relativity might just as accurately have been called the "theory of absolutes," which was in fact proposed.

The main thing that is absolute in relativity is the speed of light. The strange thing about Maxwell's discovery that electromagnetic waves travel at $3 \times 10^8$ meters per second in a vacuum is that this number, nowadays universally designated by the letter $c$, merely popped out of his equations. When we measure the velocity of a train or a baseball, it is always with respect to some other object. If we were standing in a country field we might see a train moving east at one hundred kilometers per hour with respect to the ground. From a car, however, itself moving east on a road parallel to the track at seventy-five kilometers per hour, the train appears to be moving at only twenty-five kilometers an hour. The velocity we measure of any body always depends on our *frame of reference*—roughly speaking, our vantage point or, a little more concretely, the place where we are standing.

Maxwell's result was strange because it merely says that $c = 3 \times 10^8$ meters per second. With respect to what? Maxwell himself assumed that his electromagnetic waves propagated through the *luminiferous ether*.

Water waves travel through water and sound waves travel through air, so it was natural to surmise that light waves must also travel through a medium. The luminiferous, or light-bearing, ether pervaded all space and provided an *absolute standard of rest*. If you are sitting in a train, you are at rest with respect to the train, but the train is moving relative to the earth and the earth is moving relative to the ether. Mercury also has a velocity with respect to the ether, and you can compare the earth's velocity to Mercury's by saying each has its own *absolute velocity* relative to the ether. Maxwell believed that the absolute velocity of light relative to the ether was $3 \times 10^8$ meters per second.

Unfortunately, simple calculations gave the mysterious ether rather strange properties. For instance, if the ether were one hundred times thinner than air, it must be one thousand times stiffer than diamond. More to the point, all attempts to detect it failed.

*

In 1905 Einstein took the bull by the horns and declared the ether null and void. Furthermore, he accepted Maxwell's result that the speed of light was a constant, *c;* let this be a law of nature. Thus was born Einstein's *special theory of relativity.* It is based on two simple postulates.

The first: Absolute motion does not exist. Einstein took over this axiom from Galileo and it says that no experiment done on a train can decide whether the train is at rest or moving at constant velocity. All motion is measured with respect to some frame of reference, and no reference frame is preferred over another.

The second: Any observer in any reference frame measures the speed of light in a vacuum to be $c = 3 \times 10^8$ meters per second.

A few Talmudic comments are necessary here. The first postulate is known as *the principle of relativity.* (Einstein didn't initially call his theory *relativity*; that name accrued to it over the following years, and the *theory of absolutes* was indeed proposed.) The theory is termed *special* because it concerns motion at *constant velocity*. Einstein did not address accelerated motions and assumed that the reference frames above are themselves moving at constant velocity. Motion is indeed relative in relativity.

The second postulate, apparently simple, changed everything. The idea that anyone in any reference frame measures the *same* speed of light directly contradicts Newtonian physics. If light behaves like the train passing the highway, then its velocity should depend on the reference frame of the observer, as physicists call any person or thing making a measurement.

✳

The postulate of the constancy of the speed of light also showed that space and time could no longer be thought of as separate, as they had been for centuries. It is fairly easy to see why. Imagine a clock that consists of a ball bouncing up and down in a squashed train, as illustrated at the top of page 25.

Boris on the train sees the ball merely bouncing straight up and down and can define one second to be the amount of time for the ball to make a round trip from floor to ceiling and back.

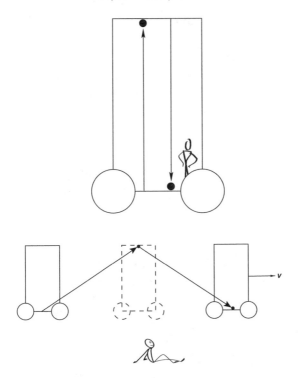

Natasha, however, observing the train from the ground as shown just above, sees the train moving to the right at its speed, *v*. One second to her remains the amount of time it takes the ball to make its round trip, but with respect to the ground the ball moves along a triangular trajectory, and thus travels farther.

But Natasha also sees the ball moving *faster*. It is bouncing vertically at the same speed Boris sees it, but to Natasha the ball is also moving forward at the train's velocity. Due to the extra speed, the ball covers the greater distance in exactly the same amount of time as measured by Boris, and one second to her is one second to him. In Newtonian physics, time is universal.

On the other hand, another of Einstein's revolutionary innovations was to realize that light is composed of particles, which for the past century have been called *photons*. If the ball is a photon, then according to relativity's second postulate, both observers measure it to have the same speed. In that case, since as seen from the ground the photon has farther to go, it must take longer to make the round trip. One second as measured by Natasha will be longer than a second as measured by Boris on the train. The discrepancy depends on the speed of the train and therefore on how far it has moved in the space of one tick.

This simple thought experiment shows that space and time measurements can no longer be thought of as independent. Einstein showed precisely how they are related, but for our purposes here those details aren't necessary. Since the advent of relativity, physicists no longer think of space and time separately; instead they speak of a four-dimensional *spacetime,* which refers to combined distances in space and time.

Although the concept of spacetime is implicit in special relativity, Einstein was not its creator. Nowhere in his early papers on relativity did he refer to time as the fourth dimension. The French mathematician Henri Poincaré saw the necessity for spacetime earlier and the German mathematician Hermann Minkowski first worked out the implications. Einstein even opposed the idea as "superfluous erudition." Ultimately, though, the spacetime viewpoint proved essential to formulating general relativity.

\*

Special relativity had other revolutionary consequences. One was that light provides the ultimate speed limit; no observer can measure a material object moving faster than light. Another is that as an object's velocity increases, its mass increases, to become infinite at $c$ (which is one reason why nothing can travel faster than light).

Yet another consequence was Einstein's immortal $E = mc^2$, which says that the energy inherent in a body is equal to its mass times the speed of light squared. By definition, however, light travels one (1) light-year per year, so in that system of units $c = 1$ and the equation says simply $E = m$. Since the advent of relativity, physicists have come to regard energy and mass as two aspects of the same thing, and they speak of "mass density" or "energy density" interchangeably, as I will.

Contrary to popular belief, Einstein was not the first person to show that mass and energy were related and, although it is heretical to say so, he never satisfactorily proved $E = mc^2$. His famous paper on the subject contains a mistake, which he attempted to patch up on subsequent occasions without success. Nevertheless, from its central role in explaining the operation of the atomic bomb or the nuclear reactions in the sun, the result has certainly withstood the test of time.

**What has been left out of special relativity?**

# GENERAL RELATIVITY, THE BASIS OF COSMOLOGY

MODERN COSMOLOGY is essentially the application of Einstein's general relativity to the entire universe. By now general relativity has become one of the most precisely tested scientific theories, if not *the* most, in history. No experiment or observation has been made that contradicts it and there is no longer any question in cosmologists' minds that the theory provides an excellent description of our universe.

While the mathematics of general relativity is complicated, its basic concepts are accessible. Before turning to the cosmos, we should try to understand how a theory called general relativity became a theory of gravity, why we believe it, and how its viewpoint shapes our concepts of space and time.

If almost all physics is about motion, then in the past several pages we have overlooked something utterly fundamental: acceleration, the change in velocity. In creating special relativity, Einstein considered objects moving at constant velocity. Nothing accelerated, and since there cannot be acceleration without a force, no forces entered the picture, either.*

Einstein intended to enlarge special relativity to include accelerations—and in doing so, he created general relativity. If general relativity is often called the most beautiful theory (which is true), it is because despite the complicated equations, the entire edifice and all its predictions spring from exactly two simple yet profound assumptions.

※

Let us begin with what Einstein called the "luckiest thought of his life." Since Galileo's day it has

---

* With some work, accelerations and forces can be put into special relativity as it stands, but this does not transform it into general relativity.

been observed that when air resistance is negligible, all objects fall to the ground at the same rate. This is the famous acceleration of gravity, usually written *g*. Near the earth's surface, *g* happens to be 9.8 meters per second per second, but the numerical value is unimportant to those of us who are not engineers. To a physicist, the important thing is that *g* does *not* depend on the mass or composition of the falling object. Gold ingots, watermelons, and feathers all fall at exactly the same rate in a vacuum.

For this reason, if we were in an elevator and the cable were cut, we'd suddenly feel weightless, because we and the elevator are falling at the same acceleration, *g*, and our feet are no longer pressing against the floor, or on the bathroom scale we have conveniently brought along.

*In a small confine, the state of free fall is indistinguishable from the absence of gravity.*

This is precisely the situation in the International Space Station: astronauts and cosmonauts fall around the earth at the same rate as the station and thus feel weightless. A more common experience is that we feel heavier than normal when accelerated upward in an elevator. In this case, gravity seems to have increased.

Einstein raised these simple observations to the status of a law of nature, which he named the *principle of equivalence:*

*In a small enough enclosure, no experiment can distinguish a constant acceleration from a uniform gravitational field.*

In other words, if the elevator is windowless it becomes impossible to determine whether the cable is accelerating us upward, or the mass of the earth has suddenly increased and hence its gravitational field. ("Gravitational field" is another way of referring to the acceleration produced by gravity, *g*.) Likewise, if the elevator cable is severed, it becomes impossible for us to know whether we are really falling toward the earth with an acceleration of *g*, or the earth has disappeared altogether. Locally, accelerations and gravitational fields are equivalent.

For this reason, Einstein understood that to enlarge special relativity to include accelerations would require a new theory of gravitation.

*

Even more than the theory of special relativity, it was his theory of gravity, going under the misleading name of general relativity, that changed our notions of space and time. The principle of equivalence alone requires that clocks at different heights in the earth's gravitational field must tick at different rates. Not only does this demonstrably happen millions of times a day, but a good deal of modern life would be impossible if it didn't.

To slightly update a thought experiment proposed by Einstein himself, imagine a rocket ship accelerating upward in empty space. Natasha, at the top of the ship, would not be caught without a cell phone in her hand. At the bottom of the ship, Boris holds an identical model. Natasha's Equivalence App sends a light flash to Boris each second according to her phone's clock. But because Boris is accelerating upward during the transit time of the flashes, he is now moving faster than he was initially and intercepting them sooner than he would have, had he continued to move at a constant velocity. He sees the pulses spaced at shorter time intervals than Natasha does and therefore concludes that his clock is running faster than hers.* If accelerations and

* Some readers may recognize that I am describing a Doppler shift.

gravitational fields are equivalent, the same must take place in the gravitational field of the earth.

The Global Positioning System relies on timing signals provided by a constellation of satellites in orbit above the earth. Because the satellites are moving at high velocity, according to special relativity their onboard clocks are ticking more slowly than cell phone clocks on the ground. Because they are in high orbit where the gravitational field is weak, general relativity also says they must be ticking more slowly than clocks on the ground. The discrepancy due to general relativity is actually twice as large as the one due to special relativity, but together they amount to something less than a billionth of a second each second.

At $3 \times 10^8$ meters per second, in one-billionth of a second light moves about a third of a meter, a foot. Unless the GPS corrected for relativistic discrepancies, each second your GPS position would get off by about one foot. Within a matter of minutes, those who no longer know how to read a map would be irretrievably lost.

General relativity is true.

\*

It also provides a description of the cosmos Newton would never have recognized. You probably know Newton's famous law of inertia, taken from Galileo, which states that a body tends to keep doing whatever it has been doing. A little more precisely, if no forces are acting on an object, it travels along a straight line. Gravity causes objects to travel along curved trajectories, as when you toss a ball and it falls to the ground. But we have just seen that in a freely falling elevator gravity disappears. In that elevator, therefore, no forces are acting on a ball and according to inertia it must follow a straight path, as on the left of the figure on page 36.

Einstein decreed that light itself behaves in the same way. Thus, in a freely falling elevator, or one moving at a constant velocity, no forces are at work and light also travels along a straight line—again, as on the left of the figure. But in an elevator accelerating upward at $g$, or above a planet with a gravitational field of $g$, the equivalence principle requires that light *must* be deflected, and by the same amount in each case, as shown at the center and right of the figure.

How strange: it seems that whether an object follows a straight or curved path depends on the

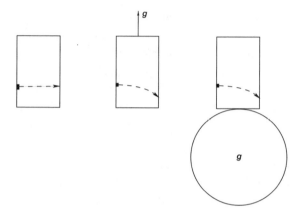

frame of reference, in the language of the previous chapter. Stranger still, it seems that whether gravity even exists depends on the frame of reference. This is true.

Imagine a building that may be built in the future, whose height is a substantial fraction of the size of the earth. At the top of such a structure, the earth's gravitational acceleration, $g$, is measurably smaller than at the bottom. This is no longer the "small confine" spoken of earlier.

If the cables are cut on two elevators, one near the building's top and the other near the bottom, they will fall at different accelerations. Someone who pitches a ball in the top elevator

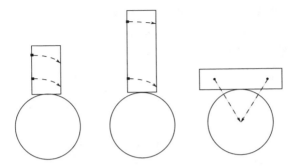

will see it move in a straight line, as will someone pitching a ball in the bottom elevator, but a person able to view both will see the balls following two different curves, which diverge. This is illustrated in the middle diagram above. Contrast this with the smaller building, on the left, where *g* is constant throughout and the two particles travel along identical trajectories, which never intersect. If the very tall building is lying on its side and two balls are dropped, they will both fall toward the center of the earth and their trajectories will eventually converge, as on the right.

This situation in which nearby particles follow identical paths but widely separated particles follow different trajectories is one of *tides.* The side of the earth nearer the sun experiences a stronger gravitational field than the opposite side. The difference in forces results in a stretching of the earth, the famous tidal bulge, as well as ocean tides.

As we have seen, we can always find a small elevator in which gravity vanishes. Tides arise when we take a more global point of view, and as on earth, tides don't go away no matter how we look at the situation. In Newtonian language, tides are really the unambiguous manifestation of gravity.

Modern cosmologists describe gravity in geometric language. On a flat piece of paper, two lines drawn parallel to each other never intersect. Indeed, this is the famous fifth postulate of Euclidean geometry. In special relativity, no forces are at work anywhere and particles moving along parallel trajectories continue to do so forever. Special relativity is the theory of flat spacetime.

On a curved surface, however, two lines that are initially parallel may eventually intersect. Two lines of longitude are parallel at earth's equator, for example, but intersect at its north and south poles, as shown on the left side of the figure on page 39. Notice also that the triangle

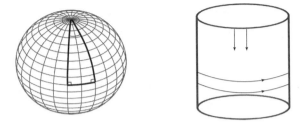

drawn on the globe contains more than 180 degrees (since the base angles alone add up to 180 degrees). This is another sign of curvature. By contrast, two parallel lines drawn on a cylinder never intersect, and so the surface of a cylinder is not curved, despite appearances.

This is precisely the situation caused by gravity. Inside an elevator, particles follow parallel lines, but more widely spaced particles follow paths characteristic of curved surfaces, which may eventually intersect. Some physicists have regarded the geometric picture of relativity as an analogy irrelevant to doing physics. The geometry of general relativity, however, is exactly the geometry of curved surfaces, developed by Georg Bernhard Riemann and others in the nineteenth century, when that is extended to include time as a fourth dimension. If it is an analogy, it is a perfect analogy. Gravity *is* the curvature of space—that is, of spacetime.

Newtonian gravity tells us that massive objects produce gravitational forces and those forces cause other objects to move. General relativity tells us that matter curves spacetime and curvature determines how other matter moves. If in the Newtonian universe forces act across a space that is forever flat, in the Einsteinian universe spacetime becomes flexible, forever changing shape as matter travels within it. This was the conceptual revolution of general relativity.

With his theory, completed in 1915, Einstein was able to exactly account for the perihelion shift of Mercury; Mercury is the innermost planet and space there is curved enough to produce a measurable discrepancy with Newtonian gravity. In 1919 a famous eclipse expedition led by Arthur Eddington showed that starlight was deflected by the gravitational field of the sun, as Einstein had predicted. A century later, general relativity has become one of the most precisely tested theories in history. That map-reading has become a lost art is living proof.

✳

General relativity, like electromagnetism, is a field theory and allows for the propagation of waves. As mentioned in Chapter 1, general relativity was not the first field theory of gravity and Einstein was not the first person to predict gravitational waves. In fact, he was initially a disbeliever, and even after coming around to their existence his first paper on the subject got it completely wrong. Nevertheless, he became the first person to get it right.

As in electromagnetism, where accelerating electrical charges produce electromagnetic waves—light or radio—in general relativity accelerating masses produce gravitational waves traveling at the speed of light. Gravitational waves are not light waves, however, and cannot be detected by ordinary telescopes. Rather, gravitational waves are tiny tidal disturbances propagating across spacetime, which stretch and shrink the measuring device itself, just as lunar tides do to earth. Because of the weakness of gravity, gravitational waves are unimaginably difficult to detect, stretching the detector an amount about ten thousand times less than the diameter of a proton. Nevertheless, after over a half-century of effort, researchers accomplished this miracle, and in

2016 the Laser Interferometer Gravitational Wave Observatory announced the discovery of gravitational waves. The wave patterns, caused by colliding black holes a billion light-years away, exactly conformed to general relativity's predictions and the discovery inaugurated a new epoch of astronomy, even as it caused tears to come to the eyes of certain cosmologists.

✳

Consequently, as far as anyone can tell, general relativity is as correct as scientific theories get. It is what physicists term a *classical* theory, meaning it takes no account of quantum mechanics. It may be necessary to create a quantum theory of gravity in order to describe the big bang *singularity,* a topic that will come up repeatedly soon enough. Barring that extreme event, however, general relativity works in every conceivable circumstance, and for that reason cosmologists do not hesitate to apply it to describe the evolution of the entire universe.

As we'll see, the real universe turns out to be nearly flat, or Euclidean, and therefore much of the formal apparatus of general relativity is nearly superfluous for modern cosmology; a Newtonian picture often suffices. Nevertheless, relativity's viewpoint is essential. In the vicinity of objects such as black holes, where the gravitational field can be extremely strong, spacetime is far from flat and there one must employ general relativity's full power.

*

Thus far I have said nothing about general relativity's second postulate. It has a rather inscrutable name, so let's just call it the "generalized" principle of relativity. Remember that special relativity concerned itself with motion at constant velocity—more precisely, with reference frames moving at constant velocity—and Einstein declared all such frames equally valid. None represented absolute space. In creating general relativity, Einstein declared that we should be able to describe motion in any reference frames whatsoever—in particular, in accelerating frames.

That declaration raises very deep questions.

Most of us have probably been on one of those amusement park rides that whirls us around in a rotating, circular cage, like a centrifuge. Indeed, we typically say that a *centrifugal force* has pushed us *out* against the cage wall. That's certainly how it feels. But a naysayer stationed on the ground would say, nay, it's a figment of our imagination. If the cage suddenly disappeared, we would fly off in a straight line as seen from the ground, in accordance with Newton's law of inertia. The centrifugal force we feel is "fictitious." In reality, the cage wall is pushing *in* on us, preventing us from flying off into space.

A spinning amusement park ride represents an *accelerating* reference frame and, according to many introductory textbooks, physics should not be practiced in such frames. The centrifugal force is fictitious because it disappears when the situation is viewed from the ground, which is not accelerating. Yet, we have already seen how gravity itself disappears in a falling elevator, which is equivalent to a nonaccelerating frame. Is gravity a fictitious force?

This question has an answer: If we believe in general relativity, we have no choice but to believe that either gravity is a fictitious force or that "fictitious forces" are real.

\*

This raises an even deeper question. We sit in a train. According to special relativity, it is impossible to determine whether it is moving with a constant velocity or at rest, but we certainly know when it begins to accelerate—we are pushed squarely back into our seats.

With respect to *what* is the train accelerating? Isaac Newton would say with respect to absolute space—the ether, which forever remains at rest. Intro physics texts agree with Newton, and in doing so are saying that the ether really does exist.

In developing his general relativity theory, Einstein was strongly influenced by the German physicist and philosopher Ernst Mach, who believed that absolute space was a figment of Newton's imagination. Given that there is no way to detect absolute space, it only makes sense to talk about accelerations relative to other material objects—for example, the stars. Einstein christened this idea "Mach's principle."

The dilemma posed by Mach had already been famously demonstrated in 1851, in Paris, when Léon Foucault set a very long pendulum swinging from the dome of the Panthéon. As the day wore on, it seemed as if the direction of the pendulum's swings slowly rotated with respect to the Panthéon's floor. In fact, the Panthéon was rotating around the pendulum, which continued swinging in the same direction with respect to the stars above. How does Foucault's pendulum "know" to swing in a direction fixed relative to the stars? Or does the reference frame of the stars coincidentally happen to be the same as absolute space? Some people don't even see a question here. Others see one of the deepest mysteries of physics.

Einstein had intended to incorporate Mach's principle into general relativity. In a universe essentially devoid of matter, one would not be able to detect any accelerations at all. To what extent Einstein succeeded in this endeavor is debated to this day, but to explore it in any depth would require another book. So I leave it there.

### How does relativity describe the entire universe?

# THE EXPANDING
# UNIVERSE

TODAY, THE IDEA that the universe is expanding is so well known that it is part of our popular culture, but what does it mean? When audience members come up to the podium after any talk on cosmology, the first question is: "If all galaxies are moving away from us, are we at the center of the universe?" and the second question is: "What is the universe expanding into?" To be honest, sometimes these questions come in reverse order, but while they are natural, they show that the concept of an expanding universe is not.

It certainly was not to Einstein. When he published the general theory of relativity in 1916, there was no astronomical evidence that the universe was expanding, and when in the same year he applied the theory to create the first modern model of the cosmos, he assumed the universe must be static. Over the next decade, astronomers were pushed to the idea of an expanding universe by the realization that nebulae—"clouds" often thought to reside within our galaxy—actually lay beyond the Milky Way; moreover, they appeared to be receding from us.

The acceptance of an expanding universe was clinched after 1929, when Edwin Hubble announced his famous "law" stating that the velocity of recession of a distant galaxy is directly proportional to its distance. For reasons that will hopefully become clear, Hubble's law implies that galaxies are receding not only from the Milky Way but from each other.*

This is exactly what astronomers mean when they speak of the expansion of the universe—galaxies are moving farther apart from one another. No discovery in cosmology has been more

* Recently, "Hubble's law" has been renamed the "Hubble-Lemâitre law," to include the Belgian priest Georges Lemâitre, who published it in 1927, but in French.

important and it lies at the foundation of the entire big bang theory. Surely, if the universe were not expanding, there could have been no big bang.

✳

Conceptually, what Hubble did was simple: he merely plotted the velocities of a number of galaxies versus their distances. Despite his data resembling the points in the figure below, Hubble, being either very brave or very foolhardy, drew a straight line through them.

Here we must confront what is, I promise, the most difficult piece of mathematics in this book: the equation for a straight line. The equation for Hubble's line is $v = Hd$, where $v$ is a galaxy's velocity, $d$ is its distance, and $H$ is the slope of the graph. The straight line implies that a gal-

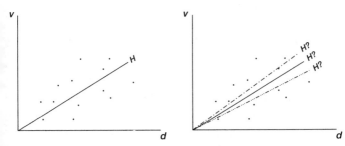

axy's recessional velocity is directly proportional to its distance: If galaxy Beta is at twice the distance as galaxy Alpha, then Beta is receding from us at twice Alpha's velocity. Moreover, the greater the slope $H$, the faster galaxies at a given distance are receding.

$H$, known as the *Hubble constant,* is easily the most famous number from cosmology and the careers of many cosmologists have been devoted to determining its exact value. Why is $H$ so important? Knowing its precise value will not likely affect the outcome of elections, but in a way we'll see shortly, $H$ measures how fast the universe is expanding, which enters into virtually every cosmological process. Furthermore, knowing $H$ gives the age of the universe, the time elapsed since the big bang. In theory, to determine $H$ is simple: following Hubble, plot galactic velocities versus their distances and read off the slope. The phrase "easier said than done" was invented for this task.

Measuring another galaxy's velocity is comparatively straightforward if we employ the famous *Doppler shift:* light frequencies from a moving object are shifted toward the red if it is moving away from us and toward the blue if it is moving toward us. Astronomers in the 1920s knew that most galaxies (or nebulae) were receding from us precisely because their light was redshifted. The exact shift depends on the object's velocity. By comparing a galaxy's observed spectrum—the frequencies of light it emits—with the known frequencies of light as measured in a laboratory, one can easily compute its recessional velocity.

The distance is the steep climb. We can't measure the distance to another galaxy with a tape measure or laser rangefinder. The distance to the nearest stars can be determined by triangulation, and the Hipparcos and Gaia satellites have extended this method to a billion stars of the Milky Way, but to measure extragalactic distances has required great ingenuity and sweat on the part of astronomers. The endeavor to establish the scale of the universe, the *cosmic distance ladder,* has probably been the major push of recent astronomy, but even with precision modern measurements, arguments over astronomical distances continue. As long as there are uncertainties in distance measurements, uncertainties will persist in almost every other astronomical quantity—in particular, in $H$.

That Hubble's own value for $H$ was about seven times the modern number hints at the difficulties involved. Looking again at the figure on page 49, it is not altogether clear that the indicated slope on the left best fits the data; other possible slopes are shown on the right. For that matter, why draw a straight line in the first place?

✳

You can better understand the implications of Hubble's law by experimentally verifying it in your kitchen. Take a wide rubber band and mark on it a series of galaxies in the form of equally spaced dots. Label them A, B, C, D. . . .Stretch the rubber band until the dots are farther apart: A . . . B . . . C . . . D.

Pretend you are located on galaxy A. If the rubber band is stretching uniformly and B moves one centimeter from A, then C has moved one centimeter from B and hence two centimeters from A. Since this all happens over the time you have been stretching the band, C must be receding from A *twice as fast* as from B.

That is Hubble's law.

The key is that the band must stretch *uniformly,* at the same rate everywhere. Any universe that expands uniformly will exhibit a Hubble law.

I said above that $H$ represents the expansion rate of the universe. Precisely, *H is the fractional rate of expansion of the universe.* In other words, *H represents the percentage increase in the distance to any galaxy per unit time.*

For example, if C is initially at five centimeters from A and it moves one centimeter in one second, then it has changed $(1/5)$ of its distance per second and $H$ is $(1/5)$ per second. The rubber band can make this clearer, but I have put the demonstration in the note below.*

Most importantly, on the rubber-band universe no particular galaxy is any more central than another. If you were located on C, then A would appear to be receding twice as fast from you as B. The picture becomes even clearer if you imagine pasting galaxies onto the surface of a balloon. As you blow up the balloon, every galaxy

---

* Suppose galaxies A and C are separated by a distance $d$, and C has a recessional velocity $v$ as measured by A. Since the rubber band obeys Hubble's law, $H = v/d$, *by definition,* velocity is the change in distance per unit time, usually written $\Delta d / \Delta t$. Thus $H = (\Delta d / d) / \Delta t$. This is the fractional change in distance per unit time.

moves away from every other galaxy, and all galaxies recede from their neighbors at the same rate. This is precisely what cosmologists mean when they speak of the universe's expansion.

So here is the answer to the first after-lecture question. Are we at the center of the universe? No.

You might well object that a balloon has a center—in its interior. Here is where the balloon analogy breaks down. A balloon is a two-dimensional surface in our three-dimensional space, and an ant on the surface can glance up into the surrounding room. The universe in which we live has three spatial dimensions and there is no surrounding room to look into. The real universe is a four-dimensional spacetime, not surrounded by anything else. The universe is growing larger, in the sense that galaxies are moving farther apart, but it is not expanding *into* anything. This is the answer to the second after-lecture question.

Of course, all this is terrifically difficult to visualize. In trying to imagine an expanding universe, people often picture in their mind's eye an expanding rubber sheet with an edge. Once we put an edge on it we are assuming an exterior, which does not exist. Once we put on an edge, we can locate a center, which also does not exist. Better is to imagine a sheet without an edge, stretching infinitely far into the distance. Galaxies marked on the sheet just keep getting farther apart from each other.

✳

At this juncture you might ask: Are galaxies themselves expanding? Are you and I expanding? No, you and I are not expanding (except perhaps through dietary habits) because electromagnetic forces are holding our bodies together. Is the solar system expanding? The usual answer is no; the gravitational attraction of the sun holds the solar system together and prevents it from expanding with the universe. Similarly, galaxies themselves are bound by gravity and do not expand.

At larger scales, things become less clear, but at approximately the scale of superclusters, which can be a billion light-years across, the force of gravity becomes insufficient to bind objects together against the universe's expansion. Only parts of superclusters may be gravitationally bound, and the superclusters as a whole may participate in the expansion of the universe. The reason superclusters are the largest structures in the universe is because anything larger would not have formed a structure at all; the universe's expansion prevents it from coalescing.

✳

Let's now run this entire chapter in reverse. If all galaxies are receding from each other, it is a fair presumption (though not a foregone conclusion) that at some moment in the past, this expansion began. The event that started off the universal expansion is what we call the *big bang,* a term coined derisively by astronomer Fred Hoyle in 1949.

The big bang was not a bang in the conventional sense; no one would have heard anything even had anyone been around to listen. It is also incorrect to imagine the big bang as a conventional explosion that took place in an already existing room. If there is no exterior to the universe, then there was no room for the universe to explode into. Spacetime as we know it came into existence at the big bang.

Finally, it is often said that at the instant of the big bang, all matter in the universe was concentrated at a single point, which must be the center. Because the universe does not have a center, this idea cannot be correct.

The rubber band can help sort this out. Assume that the band is already stretched and that A, B, C, and D are far apart. Relax the band until all the dots move back to their original position. The time for all the dots to return to their original position is the age of the universe since the big bang. Hubble's law tells us that the distance each galaxy crosses is $d = v/H$. But the distance a galaxy crosses is just its velocity multiplied by the travel time, $d = vt$, so $vt = v/H$, implying that $t = 1/H$.

The inverse of the Hubble constant is known as the *Hubble age* and is the approximate time elapsed since the big bang.

Nothing here required all the dots to be at a single location. What's more, if we imagine the rubber band to be infinitely long, with an infinite number of dots A, B, C . . . (in an infinite number of alphabets), we are required to accept that the rubber-band big bang took place everywhere along this one-dimensional surface.

It is correct to say that at the instant of the big bang all the matter in the observable universe was concentrated at a single point. The observable universe, however, is not the entire universe. The distance that light has traveled since the big bang is termed the *cosmological horizon* and, as its name implies, we cannot see anything beyond it. We are permitted to say that at the instant of the big bang everything within the cosmological horizon was concentrated at a point.

Astronomers have devised many techniques for measuring the Hubble constant that are much more sophisticated than measuring the distances to galaxies. A few of these will appear in subsequent chapters. The difficulty is that these methods do not all agree. For now, let me say only that the age of the universe—the time since the big bang—is not quite 14 billion years, or to be overly precise, 13.7 billion.

*

General relativity's prescription to describe the entire cosmos, stripped to essentials, is this: determine the contents of the universe and how they are distributed; let the equations of general relativity tell you how the universe evolves.

That may be general relativity's prescription, but it was not Einstein's. As mentioned earlier, Einstein believed the universe to be static—nonexpanding. He forced his equations to produce such a universe by adding an extra term on mathematical grounds: the infamous *cosmological constant*. It was a pure fudge factor and , once the universe's expansion was established, Einstein discarded it as "the greatest blunder of his life."

In retrospect, adding the constant seems a strange move. If a fireworks rocket exploded in outer space, the cloud of particles would initially expand rapidly, and if the fireworks particles were massive enough, the expansion of the cloud would gradually slow due to the particles' mutual gravitational attraction. Depending on the particle mass, the cloud might eventually start contracting. One thing it would never do is stand still.

In the same way, applying the equations of general relativity to the cosmos without fudging shows that it is *dynamic*. A universe without any fudge factor will automatically expand or contract at a rate determined by the density of its contents. This indeed is the primary way in which general relativity reveals the effect of gravity—in determining the expansion rate of the universe. But just as Newtonian physics does not tell us how many fireworks to load into the rocket or what should be their composition, general relativity leaves open the ingredients for any proposed universe. Once they are specified, gravity takes over and guides the evolution of the model.

In 1922, Alexander Friedmann, a Russian meteorologist, produced just such a dynamic cosmos from Einstein's equations. Because Einstein was reluctant to accept an evolving universe, it is actually Friedmann's model that has provided the mathematical basis for the big bang theory.* The important feature of Friedmann's universe is that it is as simple as a cosmological model can get. It assumes that the universe's contents are uniformly distributed and the predicted expansion is uniform—that is, happening at the same rate everywhere.

Friedmann's main equation shows exactly how the expansion rate of the universe—the Hubble "constant"—depends on its contents. The Hubble constant measured by astronomers is actually *today's* cosmological expansion rate, which is technically only a constant at the instant you read this sentence. Generally, as the universe ex-

---

* Friedmann's model was rediscovered over the years by Georges Lemâitre (1927), Howard Robertson (1935), and Arthur Walker (1936), and so cosmologists today usually refer to it as the FLRW universe.

pands the density of its contents decreases, and with it, the expansion rate.

You may remember from Chapter 3 that the matter in the universe determines the geometry of space. If the density of matter in the universe exceeds a certain *critical value,* which is about $10^{-29}$ gram per cubic centimeter (say, ten hydrogen atoms per cubic meter), then, like the massive fireworks rocket, the expansion rate in Friedmann's model will slow to zero and eventually become negative—the universe will re-collapse. Such a universe is generally referred to as *closed,* and its spatial geometry is the geometry of a spherical balloon.

If the density is less than the critical value, the universe's geometry resembles an infinitely large potato chip (on which nearby parallel lines diverge) and it will expand forever. Such a model is generally termed *open.* As mentioned in Chapter 3, the real universe seems to be flat, exactly on the border between open and closed. With an expansion rate decreasing until finally becoming zero at infinity, the universe just barely creeps toward forever.*

---

* In this discussion I am assuming that the cosmological constant is zero. If a cosmological constant is present, as is

If the expansion rate decreases toward the future, then it increases toward the past. Indeed, at the instant of the big bang it was infinite.

**Surely that is impossible?**

apparently the case in our universe (as will be discussed in Chapter 8), possible scenarios for the universe's behavior become more complicated. A spherical "closed" universe may expand forever and an "open" potato-chip universe may re-collapse.

# COSMOLOGY'S ROSETTA STONE: THE COSMIC BACKGROUND RADIATION

IF THE DISCOVERY of the expansion of the universe was the foundation of modern cosmology, then the discovery that the entire cosmos is pervaded by a uniform bath of heat at three degrees above absolute zero laid the foundation of the modern big bang theory.

A few pages ago I claimed that the universe's expansion does not necessarily imply that the cosmos started in a big bang at some definite moment in the past. The universe might have always looked more or less as it does now—in which case, as galaxies recede from one another, new galaxies must be very slowly created to fill the voids. Such a scenario was once famous as the "steady-state cosmology," according to which the universe has existed forever.

While it is difficult to imagine a universe that has existed forever, it is equally difficult to imagine a universe popping out of nothing fourteen billion years ago. Until the mid-twentieth century, there was little observational evidence to favor either the big bang or the steady-state model.

That changed almost overnight in 1965. During the previous year, two radio astronomers at Bell Labs, Arno Penzias and Robert Wilson, had been employing an extremely sensitive antenna for the Echo satellite program to investigate radio emissions from our galaxy. For accurate measurements, one must minimize any local radio interference, be it from tractor spark plugs or the apparatus itself. To their mystification, after all conceivable sources of interference had been eliminated, including bird droppings in the antenna, Penzias and Wilson discovered that an unwanted signal remained. This weak signal appeared to be absolutely the same in every direction across the sky and so could not be from the galaxy itself. Penzias telephoned

Robert Dicke, leader of the cosmology group at Princeton University, which had been readying a search for exactly this signal. After hearing him out, Dicke turned to his colleagues James Peebles and David Wilkinson and said, "Well boys, we've been scooped."

Penzias and Wilson had discovered the *cosmic microwave background radiation* (CMBR), the very heat left over from the big bang. Remaining stalwarts of the steady-state model soon enough died off and the big bang theory became the standard cosmological model. The rest of this book will trace how the standard model has evolved.

✳

What exactly is the CMBR? All hot bodies, meaning all objects at any temperature above absolute zero, emit electromagnetic energy in the form of heat. Not only do ovens and computers radiate heat, but so do rocks, fish, you, and I. For historical reasons, physicists refer to pure heat as *black body radiation* and objects that radiate it as *black bodies,* even when they aren't black.

The fundamental and remarkable property of black body radiation is that nothing about it depends on the object's composition, only on its temperature. The body's temperature tells us the amount of emitted radiation and vice versa. When a doctor's assistant points a remote-sensing thermometer at your forehead as you enter a waiting room, what is being measured is the intensity of heat radiation you are emitting and consequently your temperature, assuming that you are a black body. Penzias and Wilson applied a remote-sensing thermometer to the universe and measured its temperature, which is now known to be nearly 2.7 degrees above absolute zero.

A typical FM radio station broadcasts at a frequency of about a hundred megahertz, corresponding to a wavelength of three meters.* Unlike a radio station, a hot body broadcasts radiation across all wavelengths, but the amount radiated at each differs greatly. For a black body, the intensity of energy emitted at each wavelength—its *spectrum*—is determined by its temperature and only by its temperature. For that reason, the black body spectrum is nearly uni-

---

* One can speak interchangeably about frequency and wavelength. As frequency ($f$) goes up, wavelength ($\lambda$) goes down such that $f \times \lambda = c$, where $c$ is the speed of the wave ($3 \times 10^8$ meters per second for light, and about 340 meters per second for sound in air).

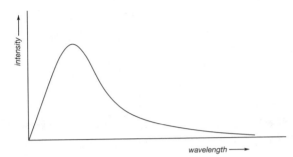

versal. It resembles the graph above, although the exact shape depends on the temperature. As you can see, most of the radiation is given off near a peak wavelength, which for a 2.7 degree black body is just about .3 centimeters, or 100 gigahertz. This is in the microwave radio band, which explains the M in CMBR.

Intensity of radiation is precisely defined as the amount of energy passing through an area of one square centimeter every second. Like the intensity of water emitted by a garden hose, it can also be thought of as the number of particles streaming through every square centimeter of space each second. Because heat is at bottom electromagnetic radiation—light—the particles in this case are photons. To say that the temperature of the CMBR is 2.7 degrees is equivalent to saying that in every cubic centimeter of intergalactic space there are currently about four hundred photons from the big bang.

Since its discovery, the CMBR spectrum has been measured by many experiments, beginning with the Cosmic Background Explorer (COBE) satellite launched in 1989, and it more perfectly matches a black body than any spectrum ever recorded in the history of civilization. In our twenty-first century, no one doubts that the CMBR represents the afterglow of the big bang.

＊

The discovery of the CMBR sounded the death knell for the steady-state cosmology because it immediately implied that the universe was *hotter* in the past than it is today. The steady-state model, in which by definition the universe was always as it is observed to be now, simply had no straightforward way to explain the CMBR's existence.

The big bang was indeed very, very, very hot. Because the universe is expanding, the density of the matter and radiation within it decreases over time; conversely, in the past the density was higher. This includes photons, which in the past were squeezed much more tightly together than they are today.

Each photon was also more energetic. As the universe expands, the wavelength of light traveling from distant regions stretches along with it, and longer wavelengths translate into redder light. This is the famous *cosmological redshift,* also often referred to as the cosmological Doppler shift, as I did in Chapter 4. To say that light becomes redder with the universal expansion is equivalent to saying that the energy of the photons making up this light is decreasing. Conversely, in the past photons were more energetic than they are now. Since temperature is simply a way of measuring photon energy, in the past photons were at a higher temperature than today. When the observable universe was two times smaller than it is now, the temperature was twice as high. It's that simple.

*

These remarks have three important consequences. The density of ordinary matter in today's universe is roughly $10^{-30}$ gram per cubic centimeter, which amounts to about one hydrogen atom per cubic meter. By contrast, via $E = mc^2$, four hundred photons per cubic centimeter, each at three degrees, constitute a mass density of about $10^{-34}$ gram per cubic centimeter. That is ten thousand times smaller than the density of matter. Discounting other ingredients, a cosmologist would say that the universe is currently *matter dominated.*

This was not always true. Going backward in time, the density of matter particles and photons increase at the same rate, like marbles squeezed together in a contracting bucket. But each photon is becoming more energetic. Thus, when the universe was about ten thousand times hotter than it is today, at some thirty thousand degrees, the energy density of photons overtook the density of matter. Before that time, which would have been about 50,000 years BB (after the big bang), the universe was *radiation dominated,* meaning that its behavior was determined by the properties of photons, not matter. The situation is sketched on the next page. The distinction between a matter dominated universe and a radiation dominated universe will become very important, very soon.

A second important, not to mention disturbing, consequence of a hot early universe is that the temperature rise does not stop. Going

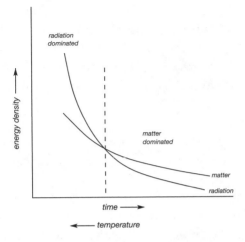

back to one second after the big bang, the temperature would have been about ten billion degrees. At the instant of the big bang itself, the temperature would have been infinite. Infinities are rarely a good sign in physics. This particular infinity, like the infinite expansion rate that put an end to Chapter 4, is a manifestation of what is known as the big bang *singularity,* which will come up ever more frequently as we close in on the big bang. If the singularity at time-equals-zero really exists, it means that the theory has completely broken down. It's much like dividing by zero—illegal. We get infinity for an answer and the equations cannot predict anything further. Usually cosmologists start their thinking about the universe slightly after the singularity, when it was presumably behaving sensibly, if not comprehensibly.

✳

Yet a third implication of a hot early universe is that the CMBR does not date from precisely the instant of the big bang.

About three-quarters of the mass of the visible universe is concentrated in the simplest element, atomic hydrogen, which consists of nothing more than an electron orbiting a proton. Because the electrons and protons carry equal and opposite electric charges, atomic hydrogen is neutral.

When the observable universe was at least a thousand times smaller than it is today, however, atomic hydrogen could not exist. The temperature was above several thousand degrees, high enough to "boil" electrons off their proton nuclei. More precisely, photons were energetic enough to knock electrons out of hydrogen atoms altogether, *ionizing* them. The resulting sea of detached electrons and protons is termed a *plasma*.

Photons cannot travel far in such a plasma because they immediately collide with the electrons and scatter. The result resembles what happens when you aim a flashlight beam through a dense fog: the beam is scattered in all directions, with the result that you cannot see far. In the early universe, as long as hydrogen was ionized, light was effectively trapped. As the temperature dropped to roughly three thousand degrees, the plasma cooled enough such that the electrons attached themselves to protons to form neutral hydrogen. Light does not interact much with neutral atoms, and after that time—which is strangely called *recombination,* although nothing was combined to begin with—light from the big bang streamed freely across the universe.

Thus, the CMBR as we observe it dates from the epoch of recombination, which modern measurements fix rather precisely at 380,000 years BB. Before this time, the universe was opaque to light, and by means of ordinary light we cannot see farther back in time than the birth of the cosmic microwave background radiation. Remember the term recombination.*

*

* The era of "recombination" is equally often called "decoupling," which emphasizes the cessation of collisions between photons and matter.

When the CMBR was first discovered, its most important feature to cosmologists was its remarkable uniformity. Its temperature, or the intensity of its radiation, as far as anyone could tell, was absolutely the same in every direction. Furthermore, on a large enough scale, galaxies themselves are distributed more or less evenly throughout the universe. Together these observations provided support for what has historically been known as the *cosmological principle:* on large enough scales, the cosmos is uniform.

The cosmological principle, buttressed by the featureless nature of the CMBR, became enshrined in the next iteration of the standard cosmological model: the model by which the universe started with a bang and this bang was absolutely uniform. No simpler picture could be imagined—but it had a number of great successes, the first of which will be discussed momentarily.

Such a simple picture could not be quite right, however, and today the most important feature of the CMBR is that it is not exactly uniform. In 1992, the COBE satellite observed slight irregularities in the temperature of the CMBR across the sky, which cosmologists knew must be there, or we wouldn't be here. These fluctuations represented the beginnings of galaxy formation. Perhaps you have seen maps of the bumps from COBE or its successors. The widely published sky map from the Planck satellite mission, launched in 2009, displays with unprecedented resolution the tiny variations in temperature of the CMBR. Although the irregularities represent a change in the background temperature of only about one hundred-thousandth of a degree, the size and distribution of the lumps has been key to unlocking almost every secret about the early universe.

*What is so important about the cosmic micro-wave background radiation?*

# THE PRIMEVAL CAULDRON

CARBON, NITROGEN, oxygen, silicon, iron...
these are elements we take for granted in daily
life, and which are necessary for life itself. It is
sobering to reflect that together such elements
comprise far less than one percent of the visible
mass of the universe. Most of the visible universe, some seventy-six percent, is made up of
the lightest chemical element, hydrogen, and
the second-lightest element, helium, makes up another twenty-four percent. Astronomy puts things
in perspective.

One of the great achievements of twentieth-
century astrophysics was the realization that

stars are nuclear furnaces, transmuting hydrogen into heavier elements, including those above. Occasionally all these elements are scattered across space by supernovae, which in the process create even heavier ones, such as lead, gold, and uranium. Ultimately, the heavy elements are incorporated into infant solar systems, planets, and us.

Essentially our entire knowledge of the composition of stars comes from observations of their spectra. The spectrum of any light source usually contains distinct lines indicating the frequencies at which chemical elements within the source are emitting light. For instance, although most terrestrial helium is created from the decay of radioactive elements deep within the earth, helium is also observed in the spectra of stars. In fact, helium, from the Greek *Helios,* was first detected in 1868 in the spectrum of our sun. Modern observations of the earliest stars indicate they were formed with masses consisting of about 24 percent helium, as well as trace amounts of other light elements.

Because the earliest stars apparently came into existence with most of their helium already present, along with a few other light elements, the question arises: How were these elements created?

In the late 1940s, physicist George Gamow and his collaborators created what is now known as the *hot* big bang theory precisely to answer this question. And answer the question it eventually did. Its success in predicting the abundances of the light elements made it, after the discoveries of the universe's expansion and the CMBR, the third early triumph of the big bang picture and one of the pillars on which the entire edifice stands.

*

The theory of the formation of the light elements in the early universe, known as big bang nucleosynthesis, or slightly more poetically, primordial nucleosynthesis, is important not only because it gives results in good accord with observations, but also because in doing so it represents a successful fusion of general relativity and nuclear physics. It also gives the first answer to the question at the end of the last chapter, on why the CMBR is essential to cosmology. Indeed, even before it was discovered, the calculations of Gamow and his colleagues assumed this cosmic heat bath must exist.

The cauldron for primordial element formation is the Friedmann universe from Chapter 4, which is assumed to have uniformly distributed contents and which is expanding at a rate determined by those contents. In its broadest outlines, the entire element-formation process is simple: start with an expanding cauldron, add the necessary ingredients, cook.

A few pages ago I persuaded you that the past universe was hotter than today's. Indeed, for a few minutes after the big bang, the universe was hot enough to permit nuclear fusion reactions, which, not unlike those that take place in the sun, processed the available ingredients into helium. As the universe expanded, its temperature dropped and, "in less time than it takes to boil a potato," as Gamow once put it, the whole process ceased. The result was 24 percent helium and the observed amounts of the other light elements.

That's the high concept, but it is neither accurate nor complete, so let us delve into a few details, where the devil lies. The most important thing to remember is that there is nothing speculative about it; the entire scenario requires only conventional physics.

To keep myself honest, technically I have been speaking of light *isotopes*. Elements are designated by the number of protons they contain;

isotopes of a particular element differ in their numbers of neutrons. An ordinary hydrogen nucleus consists of a single proton, whereas deuterium ("heavy hydrogen") is the hydrogen isotope consisting of one proton and one neutron. Ordinary helium consists of two protons and two neutrons and is called helium-4, while helium-3 consists of two protons but only one neutron.

Our goal is to produce in a very hot oven the astronomically observed abundances of these isotopes. First, the ingredients. To keep the recipe simple, we assume the material contents of the early universe to be exactly the basic building blocks that are found in the chemical elements today: neutrons, protons, and electrons. The cooking will be done by the four hundred photons per cubic centimeter (from Chapter 5) that comprise the CMBR.

There is one further ingredient: the subatomic particle called the *neutrino*. Neutrinos are the lightest of all fundamental particles, except for the photon, and they do not readily interact with other particles in nature. A single neutrino can pass through more than a light-year of lead before being stopped. For that reason, those neutrinos left over from the big bang have not yet been directly detected. One reason we know they must be present, however, is that without them the entire nucleosynthesis process could not have taken place, let alone give correct answers.

＊

That is the entire ingredients list. Next, the oven temperature must be specified. To avoid pondering the big bang singularity, when the temperature was infinite, we pick a nonzero starting time. Let us imagine the universe at .0001 second after the big bang. Projecting today's CMBR temperature of 2.7 degrees backward, we find that, at .0001 second BB, the temperature of the universe was about one trillion degrees.

To talk about such numbers may seem fantastic, but in physics a lot can happen in a ten-thousandth of a second, and a trillion degrees, while high, is not unimaginable. Ordinary protons and neutrons can exist at a trillion degrees and, moreover, the nuclear reactions among them are the ordinary ones known to physicists. At much higher temperatures, neutrons and protons would be "evaporated" into their constituents, the quarks, and no nuclear reactions could take place at all.

One trillion degrees, however, is indeed much too hot for atomic nuclei to exist. Protons and neutrons are colliding in this soup but moving too rapidly for the strong nuclear force from Chapter 1 to bind them into deuterium or helium nuclei. Just as temperatures above several thousand degrees ionize atomic hydrogen into a plasma of electrons and protons, at a trillion degrees atomic nuclei are "ionized" into a plasma of neutrons and protons.

But after about one second, the temperature has dropped to only ten billion degrees, which is roughly the temperature of the center of the sun and nearly cool enough for nuclei to begin sticking together. Assume for a moment that at one second BB there are seven protons for every neutron, as illustrated below.

At very nearly three minutes BB, the temperature has dropped to one billion degrees, which

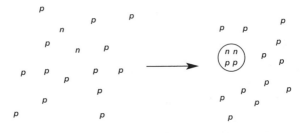

is frigid enough for colliding neutrons (n) and protons (p) to form deuterium (np). At that point, in a series of nuclear fusion reactions indeed like those found in the sun or in experimental fusion devices on earth, the deuterium is rapidly processed into helium-4, ordinary helium (ppnn).* Helium is an extremely stable element and the reactions essentially stop there. All of this takes a thousand seconds or so before everything has settled down—perhaps less time than to boil a potato, depending on the potato size.

How much helium is produced? If at the three-minute mark there are seven protons for every neutron, and all the neutrons are processed into helium, then the reactions cease once the available neutrons are exhausted. As you can see from the figure, the result is one helium nucleus for every twelve hydrogen nuclei (protons). But since a helium nucleus is four times as massive as a proton, this means that by mass we are left with

---

* The main reactions are: $n + p \rightarrow d$; $d + d \rightarrow {}^3He + n$; $d + d \rightarrow t + p$; $t + d \rightarrow {}^4He + n$;$^3$ $He + d \rightarrow {}^4He + p$; $d + d \rightarrow {}^4He$. Here $d$ represents deuterium and $t$ tritium ("extra heavy hydrogen"), which consists of a proton and two neutrons.

75 percent hydrogen and 25 percent helium, close to what is observed in the real universe.

When a computer is enlisted to do the calculations accurately, it turns out that some trace abundances of deuterium and other isotopes are left over, as in the graph below, which shows how the mass fractions of the various light isotopes evolve as the temperature of the universe drops. Getting only the helium abundance right would be a significant achievement, but remarkably, all the light-isotope abundances are in good accord with astronomical observations. This near

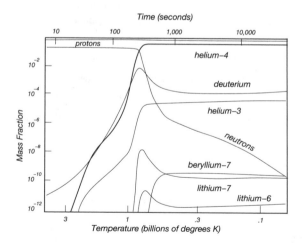

miracle is one of the main reasons that cosmologists came to believe the big bang theory.

✳

At this point, I hope you are asking where the peculiar ratio of one neutron per seven protons comes from. It is not too difficult to see how it comes about.

The first thing to realize is that neutrons and protons can be converted into one another. A neutron is essentially a proton plus an electron: $p + e \rightarrow n + v$, where the $v$ ("nu") represents a neutrino. The reaction can also proceed in reverse, converting a neutron into a proton: $n + v \rightarrow p + e$. Because these reactions are governed by the weak force mentioned in Chapter 1, they are referred to as *weak reactions,* and they show why neutrinos are an essential ingredient of nucleosynthesis.

In the early universe, because the weak reactions take place extremely rapidly, neutrons and protons are constantly being interconverted. At .0001 second BB, a proton is converted into a neutron faster than every billionth of a second. However, the neutron is slightly heavier than the proton, which means, according to $E = mc^2$, that more energy is required to create one.

Consequently, there are always fewer neutrons than protons, but higher temperatures produce more neutrons.

Imagine a bunch of billiard balls bouncing around on a billiard table, colliding with each other. The rate at which they collide depends on the number of balls, their size, and their speed, but on average there will be so many collisions per second. Now imagine that this billiard table is expanding. The bumpers are receding and so there are fewer ricochets. The table is stretching even as the balls move toward each other, resulting in fewer collisions. If the table is expanding fast enough, collisions will cease altogether.

Interesting things invariably happen in physics, and in life, when two scales cross. Large publicly funded projects may take decades to complete, but in the United States the federal government changes hands every four years; scales cross, projects are canceled, chaos ensues.

The early universe is much like an expanding billiard table. Its expansion rate depends entirely on the density of its ingredients. Projecting back from today's values shows that shortly after the big bang the density was by far dominated by photons and neutrinos. The density of neutrons and protons was so small by comparison that they played essentially no role in determining the expansion. In the words of the previous chapter, this was very much a radiation dominated universe.

At .0001 second BB, the neutron-proton interconversion governed by the weak reactions was taking place about a million times faster than the universe was expanding. As far as the weak reactions go, the universe might as well have not been expanding at all.

That situation quickly changed. As the temperature dropped, the weak reactions slowed extremely rapidly, and at about one second BB they fell below the expansion rate of the universe. Neutrinos stopped colliding with neutrons and protons and, as on the billiard table, the reactions ceased. The fraction 1:7 was the approximate ratio of neutrons to protons at this instant of "freeze-out"; thereafter, the number of neutrons didn't change too much before the onset of nucleosynthesis three minutes later.* The rest pro-

---

* Free neutrons are radioactive particles and decay with a half-life of approximately ten minutes. Thus, about twenty percent would have decayed by the start of

ceeded as already described, processing the neutrons and protons until the neutrons were exhausted at 24 percent helium.

Bear in mind that this entire discussion concerns only atomic nuclei. Atoms themselves did not form until recombination, 380,000 years later, when the temperature dropped to the point at which electrons attached to nuclei.

That the final abundance of helium is almost entirely determined by the neutron-to-proton ratio at "freeze-out" enabled cosmologists in the 1980s to predict the number of neutrino types that exist in nature well before the number was established in the laboratory. That is, known neutrinos come in three species, called *flavors,* but the possibility of more flavors could not be ruled out. The existence of any additional flavors, however, would significantly increase the expansion rate of the universe during nucleosynthesis, which would in turn increase the helium abundance (because the expansion would overtake the weak reactions earlier, at higher temperatures, when more neutrons were present).

_____

nucleosynthesis. The decay of the neutrons is charted in the graph on page 84.

Therefore, additional neutrino flavors imply a greater helium abundance. Limiting helium to the observed 24 percent ruled out new flavors, a prediction later verified in earthbound particle colliders.

*

Perhaps the most extraordinary thing about primordial nucleosynthesis, apart from the fact that it works, is that there are essentially no fudge factors. The conditions after .0001 second BB are within the realm of ordinary physics and the reactions are known from the laboratory. In the entire scenario only one number can be wiggled: the density of neutrons and protons in today's universe, which fixes their density at the time of nucleosynthesis. Since neutrons and protons are collectively known as *baryons* (for heavy particles), cosmologists speak of today's *baryon density*.

Now, stating the number of fatalities from a disease does not tell us as much as expressing it as a fraction of the population. In this case, the single input can be expressed as the ratio of photons to baryons. The photon-to-baryon ratio in our universe is roughly $10^9$ to one, a billion photons for every baryon, and it is this number that produces such good results in the nucleosyn-thesis calculations. There is no understanding, however, of why this number is $10^9$ rather than 1 or 618. Perhaps the universe merely started out with that photon-to-baryon ratio. Physicists, skeptics that they are, consider this a case of *fine-tuning*—in other words, adjusting the parameters of a model to make it fit reality. They prefer to find a natural mechanism to explain how the number arose.

"Naturally," one would expect the universe to have been created with equal amounts of matter and antimatter—there is no fundamental reason to prefer one over the other—but our universe is made almost entirely of what we term matter.* In 1967 physicist Andrei Sakharov suggested that during the big bang a slight imbalance of matter over antimatter arose—say, a billion-and-one matter particles for every billion antimatter particles. *Star Trek* fans know that matter and antimatter annihilate on contact, producing two photons per annihilation. If a billion each of

---

* Antiprotons and antielectrons, for example, have the same mass as their matter counterparts, but opposite electrical charge.

matter and antimatter particles were annihilated, one matter particle would be left over. We live in the "left over" universe, surrounded by a few billion photons per baryon. That explanation, however, only pushes the question back a notch: What determines the size of the matter-antimatter imbalance?

Although Sakharov identified the necessary conditions for the imbalance to arise, a convincing explanation for the observed photon-to-baryon ratio has been elusive. This remains an unsolved problem of physics.

In general, we do not understand how the laws of physics arose. The very success of astrophysics is convincing validation of our assumption that laws concerning momentum, conservation of energy, and so forth are the same everywhere in the universe—and cosmology's success in describing processes like primordial nucleosynthesis is convincing evidence that the natural laws have not changed significantly since the big bang.

A fundamental theorem by mathematician Emmy Noether tells us that if a system is unchanging in time, then its energy remains constant—is conserved—and if space is completely uniform, then the system's momentum (*mass × velocity*) is also conserved. But this does not explain, for instance, how space came to be uniform, and does raise the question of whether we should enlist our usual physical laws (as we will in Chapter 11) to model the universe at extremely early times, before it became uniform. What's more, when we say that "energy can be neither created nor destroyed," we are invariably referring to closed, finite systems, like breadboxes. Bread can be turned into energy, and in doing so, its mass decreases, but what it means to talk about conservation of energy for the entire universe, especially if the universe is infinite, is not well understood, if it means anything at all.

### Can we avoid fine-tuning the cosmos?

# DARK UNIVERSE

MEMBERS OF THE PUBLIC rarely ask questions about primordial nucleosynthesis after lectures. Frequently, though, comes the query: "Can you tell me what dark matter is?"

The answer should be straightforward: No.

Let us end the chapter there.

Let us reconsider. Following the dictum that Einstein never uttered about making things "as simple as possible, but no simpler," a physicist's job is to cut through nature's red tape to create the simplest explanations of observed phenomena. But nature is rarely as simple as she first appears. As observations reveal increasingly complex phenomena, the models and theories required to explain them necessarily evolve from the simple-minded to the sophisticated. Nevertheless, in contrast to economists, physicists add complications with reluctance.

With the acceptance of the big bang in the years after 1965, the standard cosmological model became the Friedmann universe with its assumption of absolutely uniform contents. But COBE's discovery of ripples in the cosmic background radiation forced a revision of the standard model to account for galaxies, galaxy clusters, and superclusters, all of which undeniably exist.

Before attacking the new standard model in Chapter 9 and Chapter 10, we must first confront the existence of *dark matter* and *dark energy,* on which the model is partly based. Perilously, the situation regarding both changes by the week. In such circumstances it is wise to enlist the *New York Times* rule: if you read about a discovery in the *New York Times* before you have heard about it from a researcher in the field, don't believe it.

※

Communication satellites circle the earth only because gravity bends their trajectories into closed orbits, counteracting the satellites' natural propensity to obey the law of inertia and fly off along straight paths into deep space. Because the force of gravity on the satellite depends on the earth's mass, the satellite's orbital velocity does as well. The higher the satellite velocity, the greater the mass required to keep it in orbit. The same applies to planets in orbit around the sun or stars orbiting the galaxy's center.

The idea of unseen matter has popped up several times over the past century and a half. In the 1930s, astronomer Fritz Zwicky noticed that the velocities of entire galaxies in galaxy clusters were much too large to be explained by the luminous mass—meaning stars—within the cluster, and he proposed the existence of *dark matter* to make up the deficit. For the moment dark matter is, well, simply matter that emits no light. Zwicky's suggestion was not taken seriously until forty years later, when Vera Rubin noticed that the velocities of stars in orbit near the edges of galaxies were also too large to be explained by the luminous matter within the galaxies. The edge stars should fly off into intergalactic space.

The measurements made by Rubin and her team were straightforward. Employing the

Doppler shift, it is easy to measure the velocities of stars circling the centers of their galaxies. By now such measurements have been performed on thousands of galaxies and clusters, and the results are invariably the same: most of the matter within galaxies is invisible. Indeed, about 85 percent of all the matter in the universe appears to be dark.

That much is nearly ironclad, and the after-lecture question is simple: What constitutes dark matter? The answer really is equally simple: We don't know. Anyone who says otherwise is either a salesman or a politician, not a scientist.

Anything that doesn't glow has been proposed as a dark matter candidate. There are so many contenders that this little book cannot discuss all of them—or indeed discuss any of them, because all candidates that have not been ruled out have not been found.

✳

Two natural thoughts for dark matter would be black holes, which by definition emit no light, and their cousins, neutron stars. Or perhaps "brown dwarfs," which are "failed stars" with masses of, say, several dozen times that of Jupiter. Brown dwarfs glow only faintly because they aren't massive enough to begin nuclear burning. Or perhaps Jupiter itself—many Jupiters—might constitute a portion of dark matter. Astronomers refer to such bodies collectively as MACHOs—Massive Astrophysical Compact Halo Objects. Unfortunately, MACHOS have been essentially ruled out as dark matter candidates, for good reason.

As discussed in Chapter 3, general relativity requires that massive bodies deflect light. Light passing around a star, a black hole, or a galaxy will be deflected from its original path, exactly as light is deflected by an ordinary lens. The result of such *gravitational lensing* is that the image of an astronomical object behind the mass-lens will shift its position or become distorted. By now gravitational lensing is a well-established phenomenon and many spectacular images have been taken by the Hubble and other modern telescopes.

Because the Milky Way is rotating, MACHOs near the edge of the galaxy rotate along with it. If light from some extragalactic source, such as an extremely bright star, passed near a MACHO acting as a gravitational lens, one would observe a slight twinkling of the star as the MACHO moved in front of it. Statistical studies of many stars in the Milky Way and Magellanic Clouds have not found any conclusive evidence for such gravitational lensing by MACHOs.

A more definite reason to exclude MACHOs is primordial nucleosynthesis. MACHOs, whatever they may be, are composed of ordinary baryonic matter (neutrons and protons), which was presumably present at nucleosynthesis times. Increasing the baryon density would increase the nuclear reaction rates forming helium during nucleosynthesis, leading to more helium. The abundance of helium actually observed by astronomers is produced when the baryon density corresponds to the *luminous* matter of the universe. If there is really five or six times more dark matter, it simply cannot reside in baryons; far too much helium would be produced during the big bang. This is a perfect example of how various aspects of a scientific theory reinforce one another.

Further, detailed analysis of the ripples in the CMBR radiation, coming in Chapter 10, requires the same ratio of dark matter to baryons as nucleosynthesis does. Whatever dark matter is, it is not the stuff we are made of.

*

That being the case, the next natural thought is neutrinos. Photons, light particles, transmit the electromagnetic force. Neutrinos are produced in situations involving the weak nuclear force and are not particles of light. They are light, however. For over half a century, in fact, physicists assumed that, like photons, neutrinos were absolutely without mass, which would of course rule them out as a dark matter candidate.

Beginning in 1998, however, that view began to change. Experiments in Japan's Super Kamiokande neutrino observatory revealed that the three neutrino flavors mentioned in Chapter 6 continually mutate into one another via oscillations. Such oscillations are analogous to the beats you hear when you hit a note on a piano and the strings are slightly out of tune. Just as the acoustic beat frequency is the difference between the frequencies of the individual notes, the rate of neutrino oscillations depends on the difference between the masses of the neutrino flavors. If the masses are zero, there are no oscillations.

Because neutrino oscillations do exist, we know that neutrinos have mass. Unfortunately, because neutrinos are such shy particles, putting an exact number on that mass has caused several decades' worth of headaches among experimental physicists. The oscillation experiments show a tiny mass difference, which suggests a similarly tiny mass, and experiments designed to detect the mass more directly indicate that a neutrino's mass must be at least half a million times smaller than an electron's mass, which is otherwise the smallest of any known particle. That in turn implies that the maximum mass for a neutrino is at least two billion times smaller than the proton or neutron mass. Measurements by the Planck satellite of the CMBR ripples suggest the neutrino mass must be smaller yet.

Consequently, in the most optimistic scenario the neutrino mass is incredibly small. But remember: there are about a billion photons for every baryon. And because neutrinos outnumber baryons by more or less the same amount (slightly less, actually) we know that, depending on the exact neutrino mass, the total mass in neutrinos could be a fraction of the mass in baryons. In the 2020s it is difficult to be certain about anything, but it seems unlikely that neutrinos can account for more than a small percentage of dark matter.

There are always "buts" in physics. In this case, there exists the possibility of a fourth species of neutrino, one that does not oscillate with the others and could have a larger mass. Such a neutrino goes by the name of *sterile*. But because the evidence for sterile neutrinos is currently inconclusive, I will leave them in peace.

\*

For several decades the leading dark-matter candidate has been, in contrast to MACHOs, WIMPs, for Weakly Interacting Massive Particles. Like neutrinos, WIMPs do not interact by the electromagnetic force—in other words, they do not emit or absorb light—so it is possible they could be dark matter. They are assumed to be massive, somewhere between ten times and a thousand times the mass of a neutron or proton, and thus they can interact with ordinary matter by gravity or by direct collisions. A weakness of the proposition is that WIMPs are completely hypothetical.

WIMP searches have been ongoing for over twenty years. Typically, a WIMP detector consists of a cryogenically cooled tank of argon or xenon gas. A WIMP collides with a xenon atom, causing it to emit a minute flash of light, which is detected by sensors surrounding the tank. The main difficulties are two. First, a WIMP is not the only particle that can engage in collisions; cosmic rays or particles from the decay of nearby radioactive elements can do the same job, and such "false positives" must be excluded. Invariably, WIMP detectors are located deep underground, usually in old mines, to screen out the unwanted background. The second difficulty is that no one really has any idea of what they are looking for, which makes it challenging to design an experiment certain to snag the culprit.

Thus far, WIMP hunts have come up empty-handed. In 2020 there was a moment of excitement when the XENON1T detector team in Italy thought it might have detected an *axion*.

Axions are regarded by many as the last best hope of dark matter. Named for a detergent, the axion was conceived in the 1970s by particle physicists to explain puzzling aspects of the strong nuclear force—specifically, why the neutron appears uniformly neutral even though its constituent particles, quarks, are charged. The axion is thought to be extremely light, even lighter than the neutrino, but in some scenarios of the early universe enough axions would be produced that they could constitute the required amount of dark matter. Such scenarios are themselves speculative, however, and since anything an author says about them is likely to end up being incorrect, let us leave it at that.

*

With so many negative results and so much conjecture, it would be surprising if scientists had not dreamed up alternative theories to compete with dark matter. To be sure, a handful of cosmologists reject the idea of dark matter altogether, suggesting instead that Newton's law of gravity be amended. At the edge of galaxies, gravity appears to be too weak to keep stars in orbit. Newton's law of gravity has not actually been tested at such large distances, however, so why not merely make it stronger out there? Such strategies are labeled MOND, for Modified Newtonian Dynamics.

One can indeed rewrite Newton's law of gravity to account for the behavior of stars at galactic edges, but this requires introducing a special length beyond which gravity is stronger than Newton would have it, and this length is equivalent to introducing a new constant of nature, analogous to the speed of light or the mass of the electron. Physicists make such moves with great reluctance. Furthermore, because Newton's law is the everyday limit of general relativity, any MOND theory requires modification of general relativity itself. Attempts to do this have been made, but at present it appears that all the attempts are inconsistent with observations. It is fair to say that most cosmologists regard MOND with far more skepticism than they do dark matter.

*

Having read this chapter, you may feel that it was less about cosmology than about particles. In a sense that is the point. The universe has proven to be an arena for exotic phenomena, and in our times one cannot divorce cosmology from the physics of elementary particles. General relativity, nuclear physics, elementary particle physics, and more have been woven together to create our present picture of the universe, and the various strands cannot be disentangled. It is well to understand that any new proposal in physics must contend with four hundred years of experiments and observations, and that inevitably nature turns out to be smarter than we are.

### Have you forgotten dark energy?

# DARKER UNIVERSE

NO, I HAVE NOT FORGOTTEN dark energy.

If it is sobering to realize that the matter of which we are composed represents only a small fraction of the matter that makes up the universe, it is even more sobering to realize that most of the universe may not be composed of matter at all. For the past twenty years, the majority of astronomers and cosmologists have accepted that most of the universe, by far, is composed of *dark energy*. The term is really nothing more than a placeholder; we have no idea what dark energy is, other than to say it is not matter and it accounts for about 70 percent of the energy content of the universe.

Perhaps this chapter should also end there, but to understand why most cosmologists believe dark energy exists, we must accept that Hubble's law, presented in Chapter 4, is a lie. That law, which states that the velocity-versus-distance graph for distant galaxies can be represented by a straight line, can be true only if the universe has been expanding at a constant rate for all time. In that case, Hubble's constant, $H$, is a genuine constant and Hubble's law holds: $v = Hd$.

On the other hand, one would naively expect that the gravitational attraction exerted by galaxies on one another should slow the universe's expansion. In that case, the most distant galaxies (whose light reaches us from early in the universe) should be receding *faster* than dictated by Hubble's law. The result is that the real graph should resemble what is shown on the next page for a decelerating universe.

In 1998, the global cosmology community was shaken, to understate matters, when two research groups, the Supernova Cosmology Project and the High-Z Supernova Research Team, independently announced that, in fact, the universe's

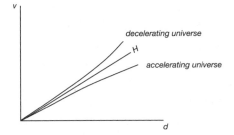

expansion was not slowing down but speeding up. The universe was evidently accelerating. Cosmologists placed bets that the results would go away, like most unbelievable results in physics, but the rival teams' evident preference to die rather than collaborate lent credence to their results. Thus far they have stood the test of time.

What the researchers did was conceptually simple. Like Hubble, they plotted the velocity versus distance for many galaxies and searched for deviations from a straight line. As in the figure, such deviations do not show up nearby and so the teams needed to measure galactic distances across a fair fraction of the observable universe.

The key to making the notoriously difficult distance measurements credible was to find a *standard candle*. As we know from life, a light bulb appears dimmer the farther away it is. Specifically, the apparent brightness of a bulb decreases with the *square* of its distance from us: if the distance doubles, the brightness goes down four times, if the distance quadruples, the brightness goes down sixteen times, and so on.

If we observe two bulbs and see that one is four times dimmer than the other, we face a dilemma: we might be observing a twenty-five-watt and a hundred-watt bulb side by side, but we could equally well be observing two hundred-watt bulbs, one twice as far away as the other. If, however, we happen to know that the two bulbs have identical wattage, then one *must* be twice as far away as the other. Furthermore, if we know that each bulb is rated at a hundred watts, that tells us exactly how much energy it is putting out. Conversely, if we measure how much energy is actually reaching us—the apparent brightness— then we know how far away the bulb is.

A standard candle is simply a light bulb for which we know the rating. In the case of the supernova projects, the standard candle was a *type 1a supernova*. A type 1a supernova is produced when a white dwarf star that has been siphoning off matter from a nearby companion collapses, releasing an enormous amount of energy. In fact, such supernovae are billions of times more luminous than our sun and for a few days one can outshine all the other stars in its parent galaxy combined, making it visible across the universe.

A survey of many type 1a supernovae led astronomers to believe that, even if they did not exactly represent a standard candle, they could be adjusted to be one, with the result that when a Hubble graph was plotted, the universe's expansion appeared to be accelerating.

＊

The acceleration implies the existence of some sort of force pushing galaxies apart. Frequently it is referred to as "anti-gravity," which is not helpful. Whatever the force is, it does not behave like gravity in reverse. For a time, the mysterious ingredient to the universe was often referred to as "quintessence"—Aristotle's fifth essence— which is an elegant term masking ignorance. More recently it has assumed the label *dark energy,* which does not explain much more, and should not be confused with the dark matter of the previous chapter. The two are not in any obvious way connected. One is matter and the other is, well, energy.

What dark energy closely resembles is Einstein's cosmological constant from Chapter 4, the fudge factor he inserted into his field equations to keep the universe static. Because Einstein used the Greek letter $\Lambda$ (lambda) to signify his fudge factor, cosmologists today often refer to dark energy as the "lambda" term in the equations. Unlike gravity, the cosmological constant really is constant, and does not change as the universe expands. Contrary to the case of a static cosmology, in our universe $\Lambda$ exerts an outward pressure that causes the expansion to accelerate.

We do not know how the cosmological constant arose. The general suspicion is that it represents the *vacuum energy* of spacetime, left over from the big bang itself. According to quantum mechanics, the vacuum of space is not empty but can be visualized as a roiling sea of energy. In physicists' minds, this sea of energy is pictured as a field of tiny, oscillating springs, which represent photons, neutrinos, and other particles. You have probably heard of the famous Heisenberg uncertainty principle, which is a law of nature. The uncertainty principle tells us that it is impossible to precisely know both the position and the velocity of a particle, or a spring, simultaneously. The energy of a spring depends on its stretch (position) and on its speed of oscillation. According to Heisenberg, these two cannot simultaneously be zero, so the vacuum springs always have some energy.

The difficulty is, if we estimate the total energy in these *zero-point oscillations* at the beginning of the universe, we find that it is at least 120 orders of magnitude larger than the dark energy today. Since that energy does not change, it remains 120 orders of magnitude larger than today's dark energy. This is the *cosmological constant problem.*

Cosmologists therefore face a choice: either $\Lambda$ is not the result of quantum fluctuations, in which case no one has the slightest idea of how it arose, or one must devise a mechanism to drive it down to the value observed today, which is about fifteen times the visible matter density. Certainly if $\Lambda$ were $10^{120}$ times larger than it is now, the universe as we know it could not exist. It would have been expanding far too rapidly for galaxies to form at all and primordial nucleosynthesis never would have taken place.

Consequently, if one believes that the cosmological constant was originally as large as simple estimates suggest, one must invent a mechanism to seriously decrease it, and very rapidly. Efforts to do so are ongoing, but there is yet no established solution.

A third choice, as usual, does exist. Recently, some cosmologists have disputed that type 1a supernovae can be used as a standard candle, suggesting that the observations are incorrect and that dark energy does not exist. That would be an elegant solution to the conundrum (if reminiscent of the flurry of excitement, in 2011, when the discovery of faster-than-light neutrinos was announced, only to have it turn out to be due to a loose connection in the equipment). Some cosmologists have other reasons for doubting dark energy, as well, but for the moment such voices are in the minority. In my hope to give this book a shelf-life longer than the time required for the ink to dry, I shall not join the debate.

Actually, there is at least one further option. If the cosmological constant were so large that galaxies could not form, then almost certainly life could not exist in that universe. The very fact that we are here asking the question argues for a small cosmological constant. This is an example of *anthropic reasoning,* to which we will return in Chapter 15.

*

You may have noticed that the cosmological constant problem is similar to the question raised by the mysterious photon-to-baryon ratio of one billion to one, which cropped up in Chapter 6. Both problems ask for an explanation of the size of a number which has no obvious reason to be what it is. You may also feel that this sort of puzzle is of a different nature than, say, attempting to determine the value of the Hubble constant, which is a purely observational issue.

That is true. The photon-to-baryon ratio and cosmological constant problems are much more *why* conundrums than *how* problems. Traditionally it has been said that science is the province of how, not why, but over the course of the past century, as the gap between observation and theory has widened, the style of theoretical physics has shifted toward why.

Such questions invariably concern what physicists term *dimensionless numbers.* As briefly noted in Chapter 6, it is always best to express quantities in ratios. To claim that a certain presidential candidate won the election by 9,870,325 votes is almost meaningless. It becomes meaningful when you discover that 9,870,325 votes is 87 percent of the ballots cast, and then you might want to challenge the outcome. A dimensionless number is merely a ratio in which the units—dimensions to physicists—have canceled out, leaving a "pure" number. The density of lead is about 11 grams per cubic centimeter, or .4 pound per cubic inch. These numbers look very different from each other and do not tell us much. On the other hand, the density of lead—in the English system, the metric system, or the potrzebie system—is about eleven times the density of water. That is a dimensionless number. Now we are comparing apples to apples, or blintzes to blintzes.

The photon-to-baryon ratio of one billion to one, and a cosmological constant 120 orders of magnitude greater than the dark energy content of the universe, are dimensionless numbers. To describe the electrostatic force between two protons as $10^{36}$ times larger than the gravitational force between two protons is to use a dimensionless number.

To ask *why* these numbers are as large as they are is to invite the response "because that's the way things are." One should not dismiss that reaction out of hand. On the other hand, physicists have it in their minds that all dimensionless numbers should "naturally" be about the same size, preferably near 1. If a particular number is orders of magnitude larger or smaller than all the others, that becomes an example of fine-tuning the universe to be what we observe it to be. Better is to find a reason that dimensionless numbers are the size they are.

In the history of physics, *why* has often enough become *how*. That many cosmologists regard the cosmological constant conundrum as the "most important problem in cosmology" shows that they take such matters seriously.

**Are fine-tuning problems real or philosophical?**

# GALAXIES EXIST
# AND SO DO WE

OTHER QUESTIONS DEMAND immediate attention. The cosmological principle described in Chapter 5 insists that the universe should be uniform when viewed on large enough scales. The caveat "large enough" is deliberately and conveniently vague, but in the name of simplicity if not philosophy, most twentieth-century cosmological calculations assumed that the universe was absolutely uniform. The primordial nucleosynthesis calculations provide a classic example. Nevertheless, the universe is not uniform. On any scale. You have probably seen computer simulations of the large-scale structure of the universe,

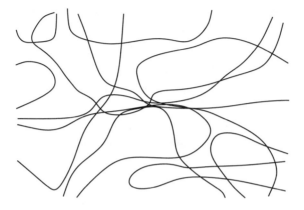

like the figure above, with long filaments resembling the interior of a lung or a Jackson Pollack painting.

The filaments are *galaxy superclusters,* the largest structures in the observable universe. Superclusters may contain hundreds of thousands of galaxies and can be hundreds of millions of light-years long. The Milky Way is so small as to be invisible in this sketch.

Because the superclusters cannot be said to be randomly distributed in any strict mathematical sense, we are confronted by the inevitable question: How did the large-scale structure of the universe come into being? If the cosmological principle were exactly true, such a web would not exist, and neither would we. The fact of an irregular universe requires that the uniform big bang model be modified into a universe that, however uniformly it may have begun, quickly became otherwise. What's more, the standard model now must become one in which ordinary matter and radiation cede place to dark matter and energy.

*

The push to understand the large-scale structure of the universe has probably been the major focus of cosmology for the past four decades. Key to the entire endeavor has been the cosmic microwave background radiation. Although for three decades after its discovery the CMBR appeared completely uniform, cosmologists knew that, for galaxies as we know them to exist, they must have begun forming at the same time the observed background was created, 380,000 years BB, and their origins must have left faint imprints on the background.

When these traces were finally discovered by COBE in 1992, the popular press—and many prominent cosmologists—went celestial, announcing the discovery of the "fingerprints of God." To be sure, champagne was cracked by the COBE team, but cosmologists knew the situation would have been more interesting had the observations revealed nothing. Physics thrives when theories and observations clash— something, somewhere is wrong. In this case, the observations simply confirmed the theoretical predictions.

The theory of galaxy formation, which I'll use as shorthand for "large-scale structure formation," may be the finest example of the unity of cosmology: It demonstrates how precision observation, particle physics, and mathematical reasoning lead to a convincing picture of our universe.

*

At its simplest level, the process of galaxy formation is one of gravity versus expansion. Gravitational attraction attempts to clump matter into structures; the universe's expansion attempts to prevent it. Who, or what, wins?

To convincingly answer this question, let us first talk about sound. And to talk about sound, let us talk about Gaul. Like all Gaul, physics is divided into three parts: particles, springs, and waves. To a physicist, what is not a particle is a spring, and anything that is not either must be a wave. Newtonian physics is the physics of particles; modern field theories are the physics of springs and waves (the discussion of vacuum energy in Chapter 8 being a pertinent illustration). A true physicist quickly reduces any problem to one about springs, waves if required, or if speaking about galaxy formation, sound waves and light springs.

A sound wave, like any wave except light, is a disturbance traveling through some medium—say, air. A stereo speaker oscillates. The speaker's oscillations alternately compress the air in front of it and allow it to expand—or rarefy, as physicists say. Indeed, a small packet of air is compressed until the air pressure within the packet has increased enough to prevent further compression, and that pressure then causes the packet air to re-expand. When the packet pressure has dropped below the pressure of the surrounding air, the ambient air compresses the packet once again. Air is a spring.

Thus, the speaker has set up a series of oscillations, which propagate across the room. It is these oscillations that form the sound wave,

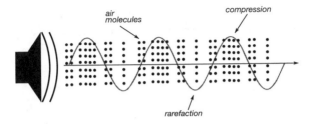

as shown in the figure above, which travels at a velocity that depends on the ambient air density and pressure. In a typical room, the speed of sound is about 340 meters per second. The stiffer the material, the higher the speed of sound. The speed of sound in steel is not quite six kilometers per second, seventeen times higher than in air.

In a simple sound wave the air pressure or density oscillates from high to low in the pattern of a classic sine wave, as in the figure. The distance between any two adjacent pressure maxima or minima is the wavelength of the disturbance, which for audible frequencies is in the meter range.*

Let us move outdoors. The earth's atmosphere is a big room, one that would collapse under its

* See footnote on page 66.

own weight if not for the air pressure supporting it against gravity. In the real atmosphere, the air pressure is quite sufficient to prevent this from happening. Just as is the case indoors, if a tall air column in the atmosphere is compressed a little, the pressure builds up and forces the column to re-expand. It overshoots until the column pressure drops below the ambient air pressure, which forces the column to re-compress. Physicists say that the atmosphere is stable against gravitational collapse and merely undergoes "acoustic oscillations"—a fancy term for sound waves.

But suppose the atmosphere were, say, a thousand times taller than the diameter of the earth. In that case, its weight would be greater than air pressure could support, and it would collapse under the force of gravity without oscillating.

*

An analogous situation existed in the early universe. If shortly after the big bang the primordial soup was spread uniformly throughout the universe, then the gravitational attraction of matter caused it to start clumping. Air pressure did not exist in the early universe, but light pressure did. In Chapter 5 we saw how, before the era of recombination, photons were unable to travel far before colliding with electrons. Photons striking matter exert a pressure on it, the same pressure that might allow sail-rigged spacecraft to cruise in the solar system under the pressure of sunlight. This pressure opposes the tendency of the matter to collapse under its own weight, and acoustic oscillations ensue, exactly as sound waves in air.

The first major difference between air in a room and light in the early universe is that the primordial soup was much stiffer than air. Steel, being stiffer than air, may have a sound velocity seventeen times higher, but the speed of sound in the early universe was nearly sixty percent the speed of light (for sticklers, $c / \sqrt{3}$). Consequently, the primordial construction material was so stiff that the *smallest* structure that could have collapsed was more massive than a supercluster of galaxies, which has a visible mass of about $10^{16}$ suns. In other words, no structures were formed in the very early universe.

Remember, though, the CMBR came into existence during recombination, when neutral atoms were formed, at which point the photons ceased striking the matter particles. That is equivalent to saying that the light pressure on the matter dropped to near zero, with the concomitant outcome that the primordial soup became much less stiff. As a result, much smaller structures could collapse—indeed, structures of about $10^5$ solar masses, which is less than a millionth the mass of the Milky Way and about the mass of a globular star cluster.

Before photons and matter parted company at recombination, they essentially acted as one soup, so when matter began to clump, photons clumped along with it. These tiny variations in photon density manifest themselves in slight temperature variations of the CMBR. It is these variations that were the fingerprints of God discovered by COBE, measured with great accuracy by its successor satellite, WMAP (Wilkinson Microwave Anisotropy Probe), and measured with extraordinary accuracy by Planck. Although the fluctuations were only about a hundred-thousandth of a degree, they were precisely large enough to produce by gravitational collapse the structures we now observe. Today, the "bottom-up" collapse scenario provides the accepted picture of galaxy formation: the smallest structures formed first, and these gradually coalesced into larger structures. Superclusters of galaxies are forming even as you read this sentence.

*Has something been left out of this picture?*

# THE UNIVERSAL
# PIPE ORGAN

THE ANALOGY of a few pages ago, comparing the universe to a room, omitted an essential difference: the universe is expanding. Because expansion pulls structures apart, it hinders gravitational collapse. The outcome of the competition depends on the exact expansion rate, which in turn depends on how much and on what ingredients are available.

Photons do not behave like matter, and dark energy does not behave like either, so it should not be too surprising that the expansion rate of the universe depends not only on the density of its contents, but also on the contents' nature. A

universe of visible or dark matter (matter dominated, in the language of Chapter 5) expands at
an ever-slowing rate. A radiation dominated
universe, where photons or neutrinos are in charge,
expands at a different ever-slowing rate. A universe filled with dark energy—ruled by the cosmological constant—increases its size at a constant
expansion rate. A highly curved universe has a
yet different behavior.

Since the expansion rate differs so much depending on the components, you might guess that
changing their proportions alters the outcome of
any galaxy-formation scenario. This is true. It's
also fortunate, because it allows cosmologists to
exclude most conceivable proposals. The question then becomes: What are the precise proportions of ingredients that permit galaxies to form
within the current age of the universe?

✳

In attempting to answer this question, let us turn
again to sound, in particular to pipe organs. The
dominant feature of a church organ is its ranks of
hundreds of pipes of differing lengths. The length
of an organ pipe determines the note it sounds.
Specifically, the pipe length determines precisely

what wavelengths or frequencies *resonate* within that pipe. Organ pipes come in many varieties, but some are essentially open at both top and bottom. As a sound wave travels through the pipe, compressing the air and allowing it to rarefy, the pressure at the open ends must remain the same as the pressure in the room. That is the condition for air to resonate within the pipe. As illustrated on pages 129 and 130, the longest wave that can be put in such a cavity that meets this requirement is one with a wavelength twice the length of the pipe. That is the fundamental, or first, harmonic—the note we hear.

A wave whose wavelength exactly equals the pipe length also meets the resonance condition. Since its wavelength is half that of the fundamental, it has twice the frequency. This is known as the first overtone, or second harmonic. The third harmonic, which oscillates at a frequency three times that of the fundamental, also resonates, and so on. In all these cases, the distance from a pressure maximum or minimum to the nearest point of room pressure is one-quarter of a wavelength, or one-quarter of an oscillation.

Basically, the universe is a pipe organ.

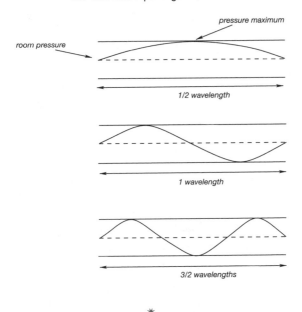

*

If we graphed the sound wave produced by an organ pipe, it would look much more complicated than a simple sine wave, but an idealized version might resemble the waveform on the left side of the figure on page 130.

Now, as you may know, a note played by an instrument consists of the fundamental plus all the overtones produced at higher frequencies.

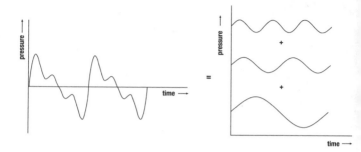

Thus we can think of any note whatsoever as being built up from the fundamental plus the overtones, as sketched on the right side above. The intensity of sound at each frequency determines the original note's shape. Mathematically, the technique used to break down a note into its overtones, or harmonics, is called *spectral analysis*. Having decomposed a wave into its harmonics, we can plot a graph like the one on page 131, which shows the amount of sound energy at each frequency. This is a sound spectrum—the same as it was for light or heat. These figures depict a simple case containing only three harmonics.

The early universe was the grandest pipe organ conceivable. Bear in mind that the temper-

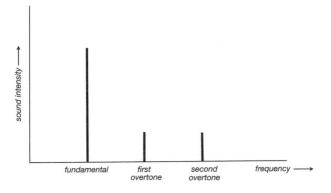

ature fluctuations detected in the cosmic microwave background are proxies for fluctuations in the matter density. These fluctuations are not all of the same magnitude. The detailed map created by the Planck space telescope shows that some fluctuations display a higher density than others, resulting in a spectrum of density fluctuations completely analogous to the sound spectrum of an organ pipe.

Indeed, the physical sizes of the density clumps are exactly determined by the resonant frequencies of the early universe. Imagine that shortly after the big bang, all matter is spread uniformly. It begins to clump, but light pressure forces the clumps to oscillate. The oscillations stop when the photons part company from matter at recombination. In the organ pipe, a maximum of pressure is one-quarter of an oscillation away from "ambient pressure," which in this case is the light pressure of the early universe. The *fundamental* oscillation in the early universe is thus the one in which a clump of matter has had a chance to compress once from its starting condition until recombination, when oscillations cease. This first overtone compresses once and expands once. The second overtone compresses once, expands once, and compresses once more.

You may object that an organ pipe has a physical *length* and here I am talking about *time*—the time between the big bang and recombination. But every time interval corresponds to a length. In this case, the length is the distance that sound traveled between the big bang and recombination. Since the speed of sound was about .6$c$, that amounts to a distance of several hundred thousand light-years. The fundamental wavelength of the fluctuations is, as in the organ pipe, four times this length. The wavelengths of the overtones are correspondingly smaller.

The universe has expanded by roughly a thousand times since these oscillations imprinted themselves on the background radiation. Because waves expand with the universe, the wavelengths of all the harmonics have stretched by the same amount, but they can be readily translated into separations as seen on today's sky. The fundamental should appear at an angular size of about one degree—twice the diameter of the moon. The overtones should appear at correspondingly smaller sizes.

Most extraordinary is that, in a series of ground-based and satellite observations spanning several decades, the predicted harmonics have been discovered. For instance, the Planck map showing the primordial density fluctuations can be decomposed into a sound spectrum. A graph of such *baryon acoustic oscillations*—sound waves to most people, fingerprints of God to enthusiasts—is shown at every cosmology seminar. As sketched on page 134, the first peak is the fundamental of the universal organ, the other peaks are the overtones.

Because clumping depends on the expansion rate of the universe, which in turn depends on its contents, this graph should reflect that. Indeed, the CMBR fluctuation spectrum has become one of the most sensitive tests of our cosmological

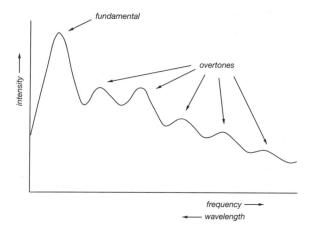

models. In a closed universe—one curved like a sphere—a distant object will appear larger than it would in flat space. This has the effect of shifting the peaks toward larger angular sizes, which on this graph is to the left. For the peaks to be exactly where they are observed, the universe must be, as far as anyone can tell, flat. This is the principal reason I stated in Chapter 3 that the geometry of the universe is nearly Euclidean—which is to say, flat.

If the universe is flat, then by definition the density of all its ingredients—ordinary matter, radiation, dark matter, dark energy—must sum to the critical density discussed in Chapter 4. That being the case, the great cosmological game is to juggle the proportions of the universe's constituents to give the best fit to the observed graph.

Take matter. If ordinary baryonic matter (neutrons and protons) were the only matter in the universe, it would have begun to clump only when the light pressure on it disappeared, at recombination. But by now you are convinced that most of the matter in the universe is dark, meaning more precisely that it does not interact with light in any way. Consequently, the light pressure of the early universe had no effect on it whatsoever, and it could not have engaged in any acoustic oscillations.

Dark matter does make its presence felt by gravity, and so it would naturally clump. Indeed, if dark matter consists of heavy WIMPs—say, one hundred times the mass of the proton—it must have begun clumping almost immediately after the big bang. Because the presence of dark matter would become appreciable at the time when the universe became matter dominated, as described in Chapter 5, which is earlier than recombination, it would have provided gravitational nucleation centers to nudge along the clumping of baryonic matter. More clumping translates into higher peaks in the primordial sound spectrum.

Suppose instead that dark matter consisted of neutrinos. Dark matter is dark matter, and in that sense neutrinos are no different than WIMPs, except that we know they exist. Neutrinos could therefore have provided the same sort of gravitational nucleation centers for baryons to give clumping a head start. The problem is that, compared to WIMPs, neutrinos are extremely light particles, streaming at nearly the speed of light in the early universe. This is much too fast to allow them to clump under their own gravity unless there were roughly a supercluster's worth of them—and in that case, the nucleation centers would be nearly the size of the universe and there would be no formation of small structures, like the globular clusters.

High-velocity particles are referred to as *hot dark matter,* in contrast to heavy, slow movers like WIMPS, known as *cold dark matter* particles. In general, the higher overtones in the acoustic spectrum, which represent clumping at smaller sizes, are washed away in hot dark matter universes. Because the higher overtones exist, cosmologists believe that dark matter in the universe is cold.

The cosmological constant, today's major ingredient in determining the expansion rate, turns out not to have a big effect on the CMBR spectrum. Although "outweighing" matter (visible and dark) in energy density *today,* it had the same energy density in the early universe—it is, after all, a constant. But the energy densities of matter and radiation rapidly increase into the past and would have overtaken the energy in the cosmological constant only a few billion years ago. Thus, the constant played little role at the formation of the CMBR, which was much earlier yet. Nevertheless, cosmologists believe it exists due to the acceleration of the universal expansion, and for other reasons I've so far left unmentioned.

One of these is *gravitational lensing of the cosmic microwave background.* Just as the MACHOs in Chapter 7 would distort the image of any light source behind them, the Planck map of the CMBR is distorted by any intervening matter—say, superclusters—lying between us and the edge of the observable universe, nearly fourteen billion light-years distant, where the CMBR was created. And just as the image produced by a magnifying glass depends on its position between the eye and the object, the distortion of the CMBR depends on the position of the lensing matter. In an expanding universe, that will depend on all the above ingredients, including the cosmological constant. Juggling the proportions to provide the best fit for the CMBR spectrum requires dark energy.

And so, at last, we arrive at today's standard cosmological model, usually abbreviated ΛCDM, for Lambda Cold Dark Matter. The best fit to the curve requires 68.5 percent dark energy, 26.7 percent dark matter, and 4.8 percent ordinary matter—but don't quote me.

*

As successful as the ΛCDM model is, it does leave open questions. First, once all the ingredients are in hand, it is reasonably straightforward to calculate the value of today's Hubble constant. Unfortunately, the value researchers find by considering baryon acoustic oscillations and gravitational lensing is about 67.4 in the standard units employed by astronomers, while the value determined by the supernova measurements is 73.9, a 10 percent discrepancy.* Astronomers pursue the Hubble constant with the zeal of cru-

---

* Astronomers would write 67.4 kilometers per second per megaparsec.

saders, and so one can be sure that they will not rest easy until the matter is resolved.

Is a 10 percent discrepancy important? Observations of small deviations from Hubble's law did lead to the discovery of the universe's acceleration. In the present situation, however, a mistake somewhere along the line is more likely. Soon enough, measurements will reach a point—say, hypothetically, where the discrepancy is one percent—when further refinements in the value of $H$ won't guide us to new physics, and it might be wise before reaching that point to ask what the aim of the pursuit is.

More importantly, I have not really talked here about structure formation, but only about the beginnings of structure formation. As the universe evolves, however, forming galaxies and stars, the physics becomes more complicated, because forces other than gravity come into play. For the record, for several hundred million years after the creation of the CMBR, the universe entered a "dark age." At the end of that period, the earliest galaxies made their appearance. Galaxies began grouping into clusters several hundred million years after that, and superclusters are still coming into existence today.

All these structures can appear within the age of the universe, assuming that the size of the fingerprints of God at the creation of the microwave background is what is observed: one part in a hundred thousand.

Moreover, the fingerprints of God spectrum has an interesting property, being what cosmologists term *scale invariant*. Loosely, scale invariant means that things look the same at any size. Zooming in on a fern leaf, you see that it appears very much the same in the small as it does in the large. Cartons of Land O'Lakes butter used to feature a Native American woman holding a Land O'Lakes carton, showing a Native American woman holding a Land O'Lakes carton, showing a Native American woman holding a Land O'Lakes carton. . . . If the sound intensity per octave in an organ-pipe spectrum never changed, we might say that the spectrum was scale invariant. If you prefer, call it the "Land O'Lakes spectrum."*

* A more accurate definition would be to say that sound intensity per cubic wavelength per octave should be constant. In the case of the CMBR, "intensity" refers to the square of the amplitude of the density fluctuations.

In the early universe, the clumping intensity compared to the clump volume remains constant. It is far from obvious that the spectrum produced by the baryon acoustic oscillations should be "Land O'Lakes," but it is.

*What fixed the size and spectrum of the fingerprints of God?*

# THE FIRST BLINK:
# COSMIC INFLATION

UP TO THIS POINT, the story has concerned the universe after .0001 second BB, when primordial nucleosynthesis was soon to get underway. It is natural to wonder what happened at earlier times, but here things become more, let's say, speculative. Going back to about a microsecond BB, we expect that neutrons and protons would be boiled into their constituent quarks, and this behavior has been recently confirmed in earth-bound particle colliders, but whether a plethora of altogether new particles makes its appearance at still earlier times is unknown. The Higgs boson would have existed in the first billionths of a second BB. The Higgs is the fabled particle that helps give mass to yet other particles, but I mention it only in passing because it does not play a central role in the cosmology plot. Clearly, thoughts of the dreaded singularity, when everything completely blows up at $t = 0$, are beginning to intrude, but for the moment let us continue to avoid a direct confrontation and ponder the first instants after the big bang, as cosmologists do, despite all the uncertainties.

Just after 1980, a new theory of the first $10^{-32}$ second BB captured the imagination of the cosmological community—and soon thereafter, the public's imagination. For reasons that will become obvious, it went by the name *inflation*, a term coined by its principal protagonist, Alan Guth, who had been giving seminars on his idea, although similar proposals had already been published by Demosthenes Kazanas in the United States and Alexei Starobinsky in the Soviet Union.

For a number of reasons, not least the name, inflation took off. Almost at once it became incorporated into the standard cosmological model, textbooks presented it as a done deal, and four decades on, inflation continues to be a cornerstone of cosmological thinking. You should understand that inflation is not a theory in the standard sense of the term, like quantum mechanics, which has been verified by myriad experiments and observations. Rather, by now, inflation represents a collection of hundreds of models whose original purpose was to explain certain "defects" in the big bang theory as I have presented it. These are not observational anomalies but theoretical or philosophical conundrums that the standard big bang simply does not address. They are much closer to the photon-to-baryon puzzle in Chapter 6 or the cosmological constant problem in Chapter 8 than they are to the perihelion shift of Mercury. Whether inflation has truly solved these mysteries has become the subject of ever more heated debate, and whether it will emerge victorious or be relegated to the ash heap of history is for future cosmologists to determine.

*

Two problems inflation was invented to solve had been long emphasized by Robert Dicke; the first of them is known as the *flatness problem*. As maintained throughout this book, the real universe, as observations confirm, is very nearly flat. Why?

"Why not?" you might respond, but the matter is not so easily dismissed. If the present universe is nearly flat, the density is close to the critical value that divides the "closed" spherical universe from the "open" potato-chip universe in Chapter 4. How likely is this? To illustrate, suppose the density today is 99.5 percent of critical. It is then easy to show that at one second after the big bang, the start of element formation, the density would have to have been within one part in $10^{17}$ of the critical value, and at $10^{-36}$ second BB, a time I have not chosen at random, it would have to have been flat to one part in $10^{52}$ or so. In other words, the universe would have to have been fine-tuned to flatness with unimaginable precision.

Even those inclined to accept the occasional coincidence find it totally improbable that the big bang could have been so flat. As with the photon-to-baryon ratio and cosmological constant conundrums, this is very much a *why* question. As before, cosmologists find it vastly preferable to transform it into a *how* question—they'd prefer to avoid any fine-tuning and find a mechanism to drive the universe to flatness, regardless of how it began.

But what does "probable" or "improbable" mean when only a single universe is at our disposal? Here we run full force into the difficulty posed by the uniqueness of the cosmos. We grapple with it in the next chapter.

\*

The second of Dicke's conundrums that inflation claimed to solve is known as the *horizon problem*. The temperature of the CMBR is observed to be remarkably uniform in all directions. Even the "fingerprints of God" of the previous chapters change the uniformity by only the thickness of a marble compared to the height of the Burj Khalifa, the world's tallest building. How did this remarkable uniformity come about? Another coincidence?

Perhaps, but to make the situation more vivid, let's say that in the observable universe there are $10^{87}$ photons, a large number. Because they are within the observable universe, they are within the distance light has traveled since the big bang— the *cosmological horizon* discussed in Chapter 4. Because no signal can travel faster than light, the cosmological horizon provides the ultimate communication barrier: no two objects can influence

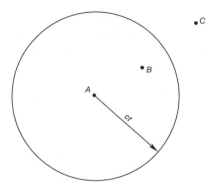

each other in any way if they lie beyond each other's horizon. As shown above, A's horizon lies at the distance light has traveled since the big bang, *(speed of light)* × *(age of universe)* =*ct*. A and B, lying within this distance, can have influenced each other. A and C cannot influence each other until the horizon has grown to the distance between them. A and B are said to be in *causal contact,* while A and C are not.

By definition, everything within today's observable universe lies within the cosmological horizon. Also by definition, the horizon grows at the speed of light, and therefore, going back toward the big bang, it shrinks at the speed of light. On the other hand, the universe's expansion rate—the rate at which galaxies are receding from one another—is less than the speed of light. Therefore, going back into the past, the universe shrinks more slowly than the horizon. Consequently, as we approach the big bang, the universe within the horizon occupies an ever-smaller fraction of what became today's observable universe. At the time the CMBR was created, only about a hundred-thousandth of today's universe lay within the horizon—say, $10^{82}$ photons.

This means that two widely separated patches of CMBR photons could not have spoken to each other at the time the background radiation was created. Like points A and C in the figure, they were not yet in causal contact. How then did they come to be at precisely the same temperature? That is the horizon problem.

\*

A third conundrum inflation claimed to solve was the *monopole problem*. According to certain *grand unified theories,* or GUTs, the strong, weak, and electromagnetic forces were unified into one grand unified "field" at the enormously high temperature of $10^{29}$ degrees that occurred at about $10^{-37}$ second BB. As the universe expanded and the unified field split into the individual fields, so-called magnetic monopoles were produced. A magnetic monopole would be an isolated north or south pole of a magnet, the magnetic analog of positive or negative electric charges. But although isolated positive and negative charges are found everywhere as protons and electrons, no one has ever observed an isolated north or south magnetic pole. All magnets have both a north and a south pole and cutting the magnet in half merely produces two smaller magnets, each with its own north and south pole.

Nevertheless, some GUTs predict that magnetic monopoles should have been produced in copious numbers in the early universe, and they would be so heavy (sixteen orders of magnitude heavier than the proton) that they would completely dominate the density of the universe. That is the monopole problem.

*

Inflation's solution to all three of these problems was elegant and straightforward enough that the average physicist could understand it. It postulated that as the GUT era ended—say, between $10^{-36}$ and $10^{-32}$ second BB—the universe underwent an enormous spurt of exponential expansion, increasing its size by twenty-seven or twenty-eight orders of magnitude in that incredibly short amount of time. This is equivalent to blowing up a popcorn kernel to the size of the observable universe.

If you were an ant walking on the surface of a popcorn kernel that suddenly inflated by twenty-seven orders of magnitude, its surface would appear exceptionally flat. This is inflation's solution to the flatness problem.

The monopole problem goes away in the same stroke. The vast numbers of monopoles in the universe were simply diluted by the enormous expansion so that their density became about one monopole per observable universe, and we haven't found it.

The horizon problem is more involved. It asks how it is that widely separated parts of the sky could have interacted and smoothed each other out to produce a uniform microwave background. Because in the standard model the horizon shrinks toward the past faster than the size of the universe, the horizon at $10^{-36}$ second was smaller than the size of the universe by about twenty-seven orders of magnitude. Thus, virtually none of the particles in the universe could interact. On the other hand, by definition, the particles within that tiny horizon could have communicated. If that patch inflated by twenty-seven orders of magnitude, it would now be the size of the observable universe.

This is what inflation claims to have done: it posits that the present universe grew out of a popcorn kernel-sized patch of sky in which the photons had already interacted and smoothed out any irregularities; inflating it would produce a uniform background radiation. Note, however, that inflation does not explain *how* the smoothing took place; it only provides the necessary condition that the smoothing *could* have occurred.

∗

A principal reason that inflation became so popular had nothing to do with these three conundrums, but with the fingerprints of God. The fluctuations in the microwave background represent temperature changes of one part in a hundred thousand compared to 2.7 degrees. They also display the Land O'Lakes spectrum, scale invariance. Both these features are observational results. How did they arise?

Early inflationary models claimed to account for them. Recall from Chapter 8 that physicists believe the vacuum of space is filled with small energy fluctuations, the so-called vacuum energy fluctuations. Inflation posits that these quantum fluctuations existed immediately after the big bang, produced in the era of quantum gravity, which will appear in Chapter 14. Inflation takes these fluctuations and, well, inflates them, until they become the fluctuations in the CMBR. What's more, it does so in such a way that the spectrum of these oscillations is Land O'Lakes.

\*

So, if inflation occurred, it could apparently explain certain puzzling features of our cosmos. But how did inflation itself come about? This is where the hundreds of different inflationary models differ. Most posit a new field, not unlike dark energy. Remember, the expansion rate of the universe depends on its contents. If the universe is dominated by dark energy—a cosmological constant—then Friedmann's equations say the size increases *exponentially* with time. In fact, because today's universe is dominated by a cosmological constant, it is now expanding exponentially, approximately.

In the inflationary scenario, much the same took place between $10^{-36}$ and $10^{-32}$ second BB. At that time, the universe was dominated by a new form of energy, which was not necessarily the dark energy of today, but resembled a cosmological constant for a time, as sketched on page 154. This nearly constant energy produced inflation's exponential expansion and, at the end of the inflationary period, decayed away until it disappeared. This figure is known as a potential energy diagram. As you may know, any system, like a ball on a hill, tends to seek the lowest energy, which is why balls roll downhill. Physicists often visualize the universe itself as a ball sitting on top of the energy curve provided by the inflationary field. As the ball slowly rolls down the almost flat hill, inflation takes place. At the end of inflation, the ball rapidly plunges into the well, losing all its energy.

Physicists, however, also subscribe to the famous law of conservation of energy and are reluctant to believe that the dominant form of energy in the universe vanished without a trace. Rather, the basic picture is this: during inflation the universe expanded enough to solve the cosmological conundrums. The enormous expansion also utterly emptied the universe of all its

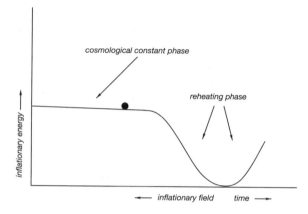

*cosmological constant phase*

*reheating phase*

inflationary energy ———→

←——— inflationary field    time ———→

contents—monopoles, photons, neutrinos, and anything else. When inflation ended, the field driving inflation decayed away, transforming its energy into the particles that make up the present universe. Inflation plus the subsequent "re-heating," as it is called, all happened in far less than the blink of an eye.

Why does the inflationary energy decay away? The original proposal was based on the well-known phenomenon of *phase transitions*. Water can be cooled far below freezing when done slowly and carefully, but if a dust particle finds its way into the water it becomes a nucleation center for ice, and the water rapidly freezes everywhere. In the context of grand unified theories, it was plausible to think that something like this happened with the vacuum energy of space in the early universe as the unified forces fragmented into their distinct identities. The vacuum energy began at a large value, then became "super-cooled," during which time inflation occurred and finally suffered a phase transition to today's value. Later versions of inflation merely posited a new field, with a potential energy diagram like the one on the previous page.

Roughly speaking, the picture sketched in this chapter is how inflation is supposed to cure the universe's headaches.

### Has something been left out?

# TO INFLATE OR
# NOT TO INFLATE

IN THE PREVIOUS DISCUSSION, I oversimplified—and even lied. While the inflationary picture provides an elegant solution to the famous cosmological puzzles, it has come under increasing scrutiny, as is proper in science, and today its future looks much less assured than it did in the years immediately following its advent.

Consider the monopole problem. Despite efforts spanning decades, no experimental evidence for GUTs has ever been found, and it may be that the theories predicting copious numbers of monopoles are simply incorrect, in which case the monopole problem vanishes.

Consider the fingerprints of God. Most accounts, popular and technical, focus on the spectrum of the fluctuations, and how that spectrum is in accord with the simplest inflationary predictions. Still, the size of the fluctuations—that hundred-thousandth of a degree—must also be explained. It has long been recognized that reproducing this number in simple models requires adjusting the shape of the potential in the diagram on page 154 to extraordinary precision— as in, change it by a part in $10^{14}$ or so and you get the wrong answer. This is another example of fine-tuning and it forces us to ask whether in choosing the potential to have the necessary shape, we have merely swapped one fine-tuning problem for another.

Furthermore, while the Land O'Lakes spectrum may be in accord with inflation's predictions, inflation is not the only process that can produce one (as we will see in the next chapter). If true, how does one decide among models? In fact, inflation actually does not predict a Land O'Lakes spectrum, only a nearly Land O'Lakes spectrum. At least a few cosmologists argue that Planck satellite data are already in conflict with inflation's true predictions and the theory should be discarded on observational grounds in favor of models to be discussed in Chapter 13. Needless to say, proponents of inflation disagree.

✳

The inflationary scenario presents cosmologists with yet other ambiguities and difficulties. For example, as has been known for over two centuries, light reflected off a windowpane is *polarized*. What does this mean? Light, an electromagnetic wave, is composed of an electric and a magnetic field oscillating at right angles to one another as the wave travels. The direction in which the electric field points is termed the direction, or axis, of polarization. Light from an incandescent bulb is *unpolarized,* meaning that the bulb emits light whose electric field is randomly pointed in all directions. Unpolarized light can be thought of as consisting of two independent light beams whose electric fields are oriented perpendicularly to one another. When such a beam strikes a window, one direction is preferentially reflected by the glass and so becomes polarized—its electric field is oscillating in one direction only.

You know this to be true. Polaroid sunglasses work because their molecules are aligned in such a way that they transmit only one direction of polarization, thus cutting the intensity of unpolarized light in half. Because light reflected off a car windshield is already polarized, if you rotate your sunglasses until their polarization axis is at right angles to the light's electric field, you see almost nothing.

The cosmic background radiation is a large car windshield. At the time the CMBR was being created, photons were striking electrons, and this set them oscillating in the direction of the light's electric field. Because shaking electrons reemit light preferentially in one direction, the light is polarized. If the primordial soup were completely uniform, photons would strike electrons equally from all directions and the overall polarization would be zero. But the tiny fingerprints of God mean that the CMBR windshield is not exactly uniform—and this results in a small, net polarization.

The polarization of the microwave background has been precisely measured by many extraordinarily sensitive telescopes—too many to list, all with acronyms like DASI and ACT, based at the South Pole or in Chile's Atacama Desert—and all verify this picture.

Now, inflation *also* predicts the existence of primordial gravitational waves, produced by fluctuating quantum fields in the very early universe. Back in Chapter 3, we encountered the gravitational waves that travel across spacetime, tidally stretching and shrinking any detector set up to measure them. They did the same to the primordial soup as the CMBR was created, producing irregularities that also polarize light. The stretching and compressing of the background by gravitational waves produce a fingerprint that differs, however, from that produced by clumping due to acoustic fluctuations (the clumping discussed in Chapter 10). In principle, with a sensitive enough telescope, the two different patterns can be distinguished.

The polarization of the CMBR due to primordial gravitational waves is predicted to be far less than that due to acoustic fluctuations, but some cosmologists maintain that if such polarization is discovered, it will provide a "smoking gun" for inflation. Despite a very public announcement at Harvard in 2014 by the BICEP2 team of just that discovery, the results were eventually retracted and to date primordial gravitational waves have gone undetected. As already mentioned, some cosmologists say that the Planck satellite data already rules inflation out.

＊

The main objections to inflation, however, spring from its fundamental assumptions. Although I have mentioned quantum fluctuations a few times already, and what inflation is supposed to do to them, it is important to understand that a quantum theory of the beginning of the universe does not yet exist. Inflation, then, cannot be a genuine quantum theory of the universe; rather, inflationary models use ordinary, classical physics to "mock up" presumed quantum behavior. Indeed, a major objection to inflation is that its fields have been introduced solely for the purpose of producing inflation, and have no observational or theoretical justification.

A related difficulty is the fact that inflation is meant to stretch the presumed primordial quantum oscillations until they become the fluctuations observed in the CMBR. No mechanism has yet been provided for the transition from the quantum to the classical theory. Indeed, if inflation proceeded a bit longer than necessary to solve the cosmological puzzles, then one can show that at the onset of inflation the wavelengths of the oscillations were less than $10^{-33}$ centimeters. This is a small number. In fact, this length, termed the *Planck length,* is the length at which physicists believe classical physics must break down altogether and below which a quantum theory of gravity must take over. Because no such theory yet exists, one must regard anything that relies on statements of what might have happened during the epoch of quantum gravity with skepticism.

For the moment, however, assume that the inflationary models sensibly reproduce quantum behavior. Quantum fields fluctuate randomly throughout the universe. Small fluctuations far outnumber large ones; nevertheless, large ones occasionally occur. During inflation, a large fluctuation in one place in the universe may move the field higher up on the curve drawn on page 154, leading to more inflation in that region before it ends. As that "bubble" inflates, more fluctuations will occur, producing daughter bubbles of longer inflation, *ad infinitum.* Inflation is eternal, literally. One therefore ends up with a very irregular situation, with inflation occurring in different amounts in different daughter universes. In some places inflation may have solved the cosmological conundrums, but in other places it has not. This *multiverse* seems to be an inevitable consequence of the inflationary paradigm, and we will consider it more fully in Chapter 15.

For the moment, the important point is that the multiverse, while extremely popular with the public, presents extreme conceptual difficulties. Suppose we tried to estimate the probability that a given universe would solve the cosmological problems. If we are dealing with an infinite number of universes, this is, to say the least, tricky. When we throw darts randomly at a dartboard that is 25 percent yellow and 75 percent black, our intuition tells us that we should hit a black sector three times as often as a yellow sector. Even faced with an infinitely large dartboard, we still feel that we should hit black three times as often as yellow, and we can indeed define probabilities in a way such that this remains true.

On the other hand, if the dartboard contains an infinite number of unique colors, then the probability of hitting any one of them is essentially zero. Suppose there are an infinite number of greens, representing all the conditions that inflation can successfully handle, but also an infinite number of reds, yellows, chartreuses, and so on. Is the probability of hitting some shade of green now greater than zero? As with the black and yellow dartboard, we would need to be able to say something like *on a finite-sized dartboard, hitting green is three times more likely than hitting purple,* and then assume this remains true even on an infinite dartboard.

Inflation presents us with this dilemma. If you ask the probability of producing a universe that solves the cosmological conundrums, you need to decide which conditions—colors—are more likely than others, and there is simply no agreed-upon way of doing that. Cosmologists Gary Gibbons and Neil Turok have concluded that most universes do not inflate enough to solve the conundrums. Mathematician Roger Penrose has gone further. The equations of inflation are exactly like those of Newton in that, if you know the present state of affairs, you can predict the future, or reconstruct the past. If you imagine a very irregular and curved universe today—one far more irregular and curved than observations permit—and run the equations back to the pre-inflationary period, you will have produced a set of conditions that, by your own construction, inflation cannot smooth out or make flat. What's more, Penrose argues that such irregular initial conditions are unimaginably more probable than smooth conditions, which leads him to conclude that inflation cannot be invoked to produce a universe resembling our own.

*

A different sort of resolution to the cosmic conundrums has frequently been proposed. One might argue that only nearly flat universes permit life to evolve. If they are too closed, they almost immediately re-collapse into a big crunch, eons before galaxies have the opportunity to form. If they are too open, galaxies are also unable to form. Therefore, out of all the possibilities resulting from the multiverse, we must observe our cosmos to be as it is, because undeniably we are here. This is another example of anthropic reasoning (about which more will be said in Chapter 15). Physicists tend to be skeptical of such arguments because there is no way to conclusively test them, but they illuminate the severe difficulties introduced by the inflationary picture, given that we have only a single universe at our disposal.

An even simpler illustration of the dilemma arises from the fact that today's universe is dominated by dark energy. If this energy really is a cosmological constant that remains constant, then as the universe continues to expand, its matter and radiation content will be diluted until only the constant remains. Even the energy provided by the curvature of space will eventually vanish—and so such a universe becomes flat. Will cosmologists of that distant epoch say there is no flatness problem, because the cosmological constant provides a mechanism to make it flat? Will they say that, because the universe's flatness depends on the size of the cosmological constant, the flatness problem is really the cosmological constant problem?

Or will all the stars in the universe by then have died out, leaving no cosmologists to ask the question?

### Are there alternatives to inflation?

# CRUNCHES AND BOUNCES

HERE, CLOSING in on $t=0$, you are asking: "What happened before the big bang?" Or perhaps: "Was there a big crunch before the big bang?" Indeed, maybe it was you who came up to the podium in the Introduction to pose this after-lecture question, one even more popular than "Are we at the center of the universe?" or "What is the universe expanding into?"

The question of what happened before the big bang is a natural one and cosmologists have been pondering it since the discovery of the expanding universe. Many have been the proposals but there is still no definitive answer. Cosmologies in which periods of expansion alternate with periods of contraction are known as cyclic universe models, or "bouncing" cosmologies, and in the past decade they have begun to be taken seriously again as alternatives to cosmic inflation.

The concept of a cyclic universe is extremely attractive because it allows us to avoid thinking about a universe suddenly popping out of nothing at a definite moment in the past. Mathematically, this means we don't need to specify the conditions at the beginning of the universe because there is no beginning. But imagining a universe that oscillates forever between expansion and contraction is not easy, either.

The difficulty faced by cyclic universes has always been the *big-bang singularity*. We can no longer put it off. In the Friedmann cosmology, at the instant of the big bang, the temperature, pressure, density, and expansion rate of the universe all become infinite. This is an utter breakdown of the system as we understand it—far more serious than a plague or economic depression, either of which eventually end. At the big bang all the equations of relativity go up in flames and we simply do not know what happened before, and perhaps never will. Friedmann himself recognized that Einstein's equations permitted an oscillating universe but paid no attention to the singularity. When in the early 1930s physicist Richard Tolman created a more detailed cyclic universe model, he recognized the severe difficulty posed by the singularity, but assumed a miracle occurred, allowing the universe to reexpand after the big crunch.

*

For decades, cosmologists believed that more irregular universes than Friedmann's might avoid the singularity. Remember, the matter in Friedmann's model is distributed uniformly and if the universe is closed, space is spherical. In a contracting universe, as in a contracting ball, all the matter approaches the looming singularity equally from all directions, eventually producing an infinite density as everything within sight is crunched together at the same time into a single point. One can, however, imagine a universe that is not so symmetrical—perhaps one shaped like a cigar. In such a universe, matter would collapse faster in one direction than another, and conceivably the singularity would be avoided.

Unfortunately, this turns out not to be the case, and all attempts made along these lines failed. The singularity remained. Essentially the failure comes about because gravity is an attractive force, which focuses matter to a point regardless of irregularities. Powerful singularity theorems by Amal Kumar Raychaudhuri, Roger Penrose, and Stephen Hawking, dating from the 1950s through 1970, prove that under fairly general conditions a big-bang singularity is unavoidable.

But all theorems make assumptions, and the big-bang singularity can be evaded by introducing a large enough repulsive force. The cosmological constant—dark energy—accelerates galaxies away from each other, providing exactly the sort of repulsive force necessary to dodge the singularity. The main questions are these: How big should it be to produce a big bounce without interfering with astronomical observations? And should it really be constant?

For instance, suppose the current expansion of our universe was preceded by a collapse. In the collapsing phase, the CMBR would be heating up and one might postulate a cosmological constant large enough to bounce the universe before it reached a temperature of one billion degrees, which would take place three minutes before the big crunch. However, after the bounce—our bang—no primordial nucleosynthesis would take place and, unless the light isotopes already existed in their current abundances, they would never be created. What's more, such a large cosmological constant would cause the universe to expand so rapidly that galaxies wouldn't form. Adding a simple cosmological constant to cure the Friedmann model of its singularity is not a viable option.

The trick, then, is to introduce something that resembles a cosmological constant at the beginning of the universe—perhaps like the potential energy diagrammed on page 154—which then disappears before it causes havoc. Numerous proposals have been made, all differing in their features and motivations, and we will not go into the gory details. An attractive option is to bounce the universe before it contracts to the Planck size of $10^{-33}$ centimeter, mentioned in Chapter 12, which would be reached at the Planck time, $10^{-43}$ second before the big crunch.

The Planck length and time mark the end of physics as we know it. At smaller lengths and shorter times, our usual conceptions of space and time probably break down altogether and a theory of quantum gravity is presumed necessary to describe the singularity or get through it. Quantum mechanics can indeed produce repulsive forces that might do the job, but as already mentioned, a theory of quantum gravity does not exist. If instead, a bounce occurs well before the Planck scales are reached, then there is no need to invoke quantum mechanics. In that case, we can rely solely on conventional physics, which does exist.

*

In the past decade, some bouncing cosmologies have exploited these precepts. Like inflation, they invoke a new field resembling a cosmological constant that causes a bounce, but in which the blessed event takes place at a time of about $10^{-35}$ second BB. That is a long time (in physicists' minds) before the Planck era is reached; it is even before the GUT era is attained, in which case classical physics should be entirely adequate.

You should be wondering whether such models can solve the cosmological conundrums that inflation was designed to explain away. As it happens, some of them can, and in much the same way.

To understand how, first realize that the instant explanation I gave in Chapter 11 for inflation's solution to the flatness problem—that the universe merely inflated twenty-seven orders of magnitude in a blink to make it appear flat—was a lie (although one commonly perpetrated by cosmologists). If we stand on the beach, looking out over the ocean, the earth appears flat to us precisely because our horizon is only a few kilometers away, which is far smaller than the size of the earth. But if we were standing atop a mountain whose height was comparable to the radius of the earth, we would clearly see the earth's curvature.

So flatness is relative; you must always compare the distance to the horizon with the size of the earth. If the horizon is much smaller than the radius of the earth, the earth appears flat. Similarly, in Chapter 11 we saw that in a collapsing cosmos the horizon always shrinks faster than the universe does, so the universe looks ever flatter toward the big bang.

The same applies in bouncing cosmologies. As we approach the big crunch in a collapsing universe, the universe appears flatter and flatter because we see only smaller and smaller distances. It is this little, flat piece of spacetime territory that becomes our present universe after the bounce.

The horizon problem goes away in the same stroke. If you imagine the universe in the dim past, just as it began to re-collapse in the previous cycle, all parts of that universe are already able to communicate because they lie within the horizon. As the universe shrinks toward the crunch, the horizon shrinks faster and it is the small patch within the horizon that becomes the present universe after the bounce, as it did in inflation. Since all particles in the patch already communicated before the bounce, there is no longer a horizon problem.

One striking feature of modern bouncing cosmologies is that these problems can be solved by a very slow contraction, such that the collapsing phase does not necessarily mirror the expanding phase in reverse. In some models, the universe does not even have to contract much to do the job. Moreover, as hinted in the previous chapter, an exponential expansion is not the only mechanism that can produce the Land O'Lakes spectrum in the microwave background. Mathematically, the slow contraction of some models does exactly the same thing.

Also, do not forget that the primordial gravitational waves predicted by inflation but not yet discovered are assumed to be the result of fluctuations created during the epoch of quantum gravity. Because in bouncing cosmologies that epoch is never attained, essentially no primordial gravitational waves are produced. The multiverse, the unruly offspring of those quantum fluctuations, is not produced either.

Bouncing cosmologies are currently an active field of research, but history teaches us that active areas of research may find themselves abandoned in the blink of an eye. So, while it is early to decide whether a big bounce will cure the conceptual headaches induced by inflation, in this blink of an eye they do appear to be an attractive and viable alternative.

**How does one know whether such theories are true?**

# WHY QUANTUM GRAVITY?

WE HAVE ARRIVED at $10^{-43}$ second after the big bang. It is time—if time means anything—to create a theory of quantum gravity. Should bouncing universes turn out to be unviable in avoiding the singularity, cosmologists will have no other option. The main drive to create a theory of quantum gravity, however, is not so much the singularity itself as physicists' centuries-old conviction that the forces of nature should be unified into one towering edifice, the legendary *unified field theory*.

No observation ever made contradicts general relativity, and it is therefore considered to be as correct as a scientific theory gets. Yet it is a classical theory, taking no account of quantum phenomena. Modern quantum field theories have been tested to about the same precision as general relativity—arguments persist over the winner—but they take no account of gravity.

Theoretical physicists are convinced to the marrow of their bones that these two very different species should be joined into a consistent quantum theory of gravity. Nearly a century of effort, however, has gone into arranging a marriage without success. On the coarsest level, the difficulty has been that general relativity is a theory of the very large, while quantum theory is a theory of the very small. That clarification is unlikely to satisfy, but as physicist John Wheeler once remarked, the most difficult question about quantum gravity is: What is the question?

Let us ask a few basic questions, then; expect no answers.

First, what are quantum phenomenon? And at what point should quantum mechanics and relativity be wedded? The word *quantum* has long been part of our popular vocabulary, but despite the efforts of automobile branders and quantum healers, its exact meaning may remain fuzzy. In classical physics, most properties of a system—its energy for example—are permitted in any amount. The basic precept of quantum mechanics is that, no, these quantities come in discrete, or *quantized,* units, just as cash comes only in integral multiples of pennies. When Max Planck created quantum mechanics in 1900 by explaining the very black body spectrum of Chapter 5, his fundamental postulate was that the light emitted by the black body was quantized such that its energy equaled only integer amounts of the light frequency multiplied by a new constant of nature, which he labeled $h$. This number, now universally called *Planck's constant,* fixes the size of all quantum phenomenon.

In 1905, Einstein showed that not only was light quantized in Planck's sense, but that light should actually be associated with packets of energy, or *quanta,* which behave as particles. When Planck talked about the black body emitting light, he really meant light quanta—photons. A photon's energy is given by the light frequency multiplied by $h$. Swarms of photons acting in concert constitute a light wave, and when we study waves we no longer pay attention to the properties of individual quanta. A light wave is described by Maxwell's classical theory of electromagnetism.

One way of saying that a theory is quantum is that $h$ is in there somewhere. If a theory doesn't contain $h$ it is a classical theory. You won't find $h$ in general relativity no matter how hard you look. On the other hand, being a classical theory of gravity, every one of its equations features Newton's gravitational constant $G$, which determines the strength of the gravitational force.*

The second important feature of quantum mechanics involves a famous phrase: *wave-particle duality*. Just as light can behave as particles, particles can behave as waves. Every particle has wave properties associated with it. In particular, it has a wavelength, which depends on the particle's mass and its velocity—and on $h$. Think of this wavelength as the quantum size of the particle, its size when it is behaving like a wave. For subatomic particles, like electrons, the wavelengths tend to be very small, roughly the size of an atom, and are unnoticeable in everyday life. In systems of atomic size, however, as inside modern electronics,

* See footnote, page 14.

the wave nature of matter becomes extremely important.

✳

With these concepts we can understand the scales at which general relativity and quantum mechanics should be joined—precisely, the Planck mass and the Planck time of the previous chapters. You may know that any unit of measurement, be it metric, English, or potrzebie, is based on three fundamental quantities: mass, length, and time. The question is, what is the most sensible way to choose these three basic quantities?

In the nineteenth century, physicist George J. Stoney argued that it was better to base units of measurement on naturally occurring quantities, such as the electron's charge, the speed of light $c$, and the gravitational constant $G$, rather than on the length of a stick in Paris. Later, Max Planck had the same thought and proposed that the fundamental constants $G$, $h$, and $c$ be made the basis for a system of units, today called natural, or Planckian, units. With a little patience you can combine $G$, $h$, and $c$ into a length, which is about

$10^{-33}$ centimeter, a time, which is about $10^{-43}$ second, and a mass, which is about $10^{-5}$ gram.*

Clearly, the Planck length and time are unimaginably smaller than anything you (or most physicists) would ever contemplate, while the Planck mass is unimaginably large compared to the mass of subatomic particles—large enough to be measured on a modern balance. If you multiply the Planck mass by $c^2$, you get the Planck energy, which is about $10^{15}$ times higher than the energies produced in the Large Hadron Collider, the most energetic particle accelerator on earth.

What do these bizarre numbers signify? The fundamental constants are the most important numbers in the universe because they determine the domain of all natural phenomena. $G$ sets the strength of the gravitational force, while $h$ determines when quantum effects are significant. When $c$ appears in a situation, it shows that relativity is important—something is moving near the speed of light.

---

* Specifically, the Planck mass is $m_p = \sqrt{hc/G}$; the Planck length is $\ell_p = \sqrt{hG/c^3}$; the Planck time is $t_p = \sqrt{hG/c^5}$.

You probably know that a black hole is an object whose gravitational field is so strong that light cannot escape; its size is given by its mass and $G$ and $c$, nothing else. The size of a black hole can be thought of as the scale on which gravitational effects become extremely important. If you ask for the mass of a particle whose quantum size—its wavelength—is the same as its gravitational size, you get the Planck mass. The size of this quantum black hole is the Planck length, and the time for light to cross it is the Planck time.

So, the Planck scales represent the lengths, times, and energies at which quantum effects and gravitational effects are equally important. At these scales we cannot ignore either gravity or quantum mechanics and must create a quantum theory of gravity to describe the universe.

※

Why has such a theory proved so difficult to create? Fundamentally it is because the basic assumptions of general relativity and quantum mechanics are so different. Quantum mechanics ignores gravity and general relativity ignores quantum mechanics. Put another way, quantum theories assume spacetime is always flat, as in special relativity. General relativity assumes that spacetime can be curved, depending on its matter content.

This is a serious problem, which results in extraordinary technical difficulties. As originally created, quantum mechanics was, like Newtonian physics, a theory of particles. And like Newtonian mechanics, it took no account of special relativity. Wedding quantum mechanics and special relativity into *relativistic quantum mechanics* was accomplished by Paul Dirac in the late 1920s.

Relativistic quantum mechanics, however, continued to concern itself with particles—in particular, with electrons, which are regarded as point particles. Points, by definition, have zero extent. This produces the serious difficulty that when two point electrons touch each other, the electrical force between them becomes infinite.* Similarly, the energy of a point electron's field becomes infinite as one approaches the electron,

---

* The electrical force between two point particles looks just like the law of gravity (see footnote, page 14), except that the masses are replaced by the electric charges and $G$ is replaced by another constant. As the distance $r$ between the two particles goes to zero, the force becomes infinite.

and therefore so does its mass, which by $E = mc^2$ must include the energy of the field.

The efforts to resolve these dilemmas led to quantum field theories. In particular, *quantum electrodynamics* became the theory that explained how electrons interacted with photons. The naive hope was that, by smearing things out into fields, we need never get too close to point electrons, and such infinities—such singularities—would disappear.

A little less vaguely, in quantum field theory all interactions are described by exchanges of particles—the electromagnetic force is really due to an exchange of photons. Such exchange particles are termed *virtual*. We can regard them as manifestations of the vacuum fluctuations discussed in Chapter 8. According to the uncertainty principle, because the energy of the vacuum is fluctuating and never exactly zero, it can spontaneously create particles so long as they do not live longer than the uncertainty principle permits; this is why they are termed virtual. The expectation was that surrounding a point electron with a cloud of virtual particles would soften the singularities.

Vain hope. Matters got worse and infinities appeared everywhere. Mathematical methods known as *renormalization* were invented to cure the theory of infinities and give finite answers— which miraculously agree with experiment to such a precision that quantum electrodynamics is often called the most precisely tested theory ever created.

Originally, no one understood why renormalization worked. Even one of its inventors, Richard Feynman, called it "hocus-pocus." Nowadays, the process is on firmer mathematical footing, but in any case renormalization is still considered essential for a viable field theory; if a theory cannot be renormalized to give sensible answers, it is discarded.

Unfortunately, not only do the infinities persist in standard attempts to quantize gravity, but the renormalization process fails and the theory cannot give sensible results.

*

That grave difficulty has resulted in a profusion of approaches toward creating a full theory of quantum gravity. The simplest avenue is to assume that gravity can be described classically, by general relativity, while treating any other fields in the problem, such as light, by the methods of quantum field theory. Physicists refer to such an approach as "semi-classical," which is a polite way of calling it a bastardized tactic. Nevertheless, it can be expected to bear fruit when the gravitational fields in the problem are not *too* strong—say, around large enough black holes. (The larger the black hole, the weaker its field.) For sure, the semi-classical approach resulted in quantum gravity's most famous triumph: Stephen Hawking followed this route to his celebrated 1974 discovery that black holes are not completely black but radiate energy, exactly the heat of black bodies.

Because it is so weak, black hole radiation has not been directly observed. That the temperature of a black hole of one solar mass would be about a ten-millionth of a degree, and the temperature of larger black holes even less, gives an idea of its feebleness. The fact, however, that Hawking's calculation showed that the radiation should be precisely that of a black body led most physicists to immediately accept the amazing result.

If black holes radiate energy, they must be losing mass. As they lose mass, their temperature increases; they emit energy more rapidly, losing mass more rapidly. This runaway effect led

Hawking to predict that black holes would eventually end their lives in spectacular explosions. But his method actually assumes that the gravitational field, and thus the mass of the black hole, do not decrease. Such predictions, therefore, must be considered somewhat speculative. Indeed, the evaporation process should exert a feedback on the black hole such as to slow further evaporation; at least one of Hawking's colleagues claims, in fact, to have demonstrated that the feedback halts the evaporation long before any explosion takes place.

That result may turn out to be incorrect, but the example illustrates how difficult the issues are and how far we are from a full theory of quantum gravity. It is clear that Hawking's approach cannot be applicable at the Planck time.

*

What might be? Applicable at the Planck time, that is?

The most famous attack on the problem has been *string theory,* which lies beyond the scope of this little book. String theory attempts to be a unified field theory, or what is popularly called a *theory of everything*—a theory that not only unites the electromagnetic and nuclear forces (as do GUTS)—but includes gravity as well. String theory is a quantum field theory, but one in which the fundamental building blocks are not point particles; instead, they are tiny strings, whose length is approximately the Planck length. Once again, smearing points into finite strings might expunge infinities. The strings can have either open ends flapping about or be closed into loops. Ordinary particles are viewed as overtones of string vibrations, in the same way a violin string (or organ pipe) produces overtones.

A major difference between the strings of string theory and ordinary strings is that ordinary strings live in our universe of four spacetime dimensions (one time and three space), while, in one version of string theory, strings live in universes with ten spacetime dimensions (one time and nine space). The extra spatial dimensions are assumed to curl up on themselves, as around a cylinder, on lengths comparable to the Planck length. This is small enough so that we don't notice them.

String theory has had a number of mathematical successes. The most celebrated is that theorists have used it to derive the famous entropy of black holes, proposed by Jacob Becken-stein and made more precise by Hawking. (I won't talk about black hole entropy, but the result is famous and intimately related to the idea that black holes have a temperature.) String theory also predicts the particle that exchanges the gravita-tional force—the *graviton,* about which I'll say a little more shortly.

The appearance of the Planck length in string theory immediately tells us that it should indeed be a theory describing the extremely early uni-verse. That is actually a severe difficulty; so far, string theory has made very little contact with other branches of physics. In particular, no earth-bound experiment has been able to lend it any confirmation. What's more, the ten-dimension version is based on the concept from particle physics known as *supersymmetry,* which unites matter particles (like protons) with force parti-cles (like photons) into a larger group. Not only is there no experimental evidence for supersym-metry, but results from the Large Hadron Col-lider seem to have all but ruled out the simplest versions.

Furthermore, the original attraction of super-string theory was that only one version of the theory appeared to be mathematically consistent. Nowadays, however, it is conceded that there may be $10^{500}$ different versions, a rather large proliferation of possibilities known as the *string-theory landscape*. The landscape should remind you of the multiverse from Chapter 12. One might reasonably argue that any theory that produces $10^{500}$ universes has not predicted anything. This is a serious issue.

\*

Another attack on quantum gravity, not quite so well known as string theory, is *loop quantum gravity*. It does not intend to be a theory of everything but confines itself to quantizing gravity. It bears some resemblance to string theory in that its basic entities are loops, about the Planck length in size—but loop-gravity loops are four-dimensional. Indeed, they may be viewed not as existing in spacetime but as providing the basic building blocks of spacetime. Loop gravity calculations have also reproduced the Beckenstein-Hawking entropy of black holes.

In loop gravity, it simply does not make sense to talk about lengths smaller than the Planck length and times shorter than the Planck time;

space and time themselves are quantized. It may help to visualize spacetime as a flexible lattice, whose bendable struts are of the Planck length and time. More closely, it probably resembles what, since long before the advent of loop gravity, has been popularly called *quantum foam*.

I have not emphasized the third important respect in which quantum mechanics differs from Newtonian physics, an aspect that goes hand in hand with the uncertainty principle. Quantum mechanics is a *probabilistic* theory. Unlike Newtonian mechanics, which tells us exactly where a particle will be in the future if we know its present position and velocity, quantum mechanics tells us only the probability that it will be in a certain place at a certain time.

It may be, then, that nothing so definable as "one centimeter" or "one second" exists in the Planck era. Quantum foam will require some probabilistic description that only "crystalizes" into our universe once the Planck era ends.

How would a quantum theory of gravity avert the singularity? Quantum fluctuations produce a pressure that manifests itself much like the repulsive force of the cosmological constant. If large enough it can bounce the universe during the

Planck epoch. The exact results depend on the particular model being considered, which are too many to count. Loop quantum gravity claims to be able to do this, but no theory of quantum gravity has solved the cosmological constant problem—why today's cosmological constant is as small as it is.

One thing is nearly certain: to resemble our conventional field theories, in which forces are transmitted by particles, any theory of quantum gravity should predict the existence of a graviton, which would transmit the gravitational force. String theory does this. Although gravitational waves have been detected, however, individual gravitons have not and very likely never will be. If neutrinos interact with ordinary matter so rarely that one can pass through light-years of lead before hitting anything, then a graviton would interact with matter about twenty orders of magnitude *less* frequently, making direct detection of gravitons almost inconceivable.

This raises questions about how one could experimentally test a quantum theory of gravity. Some physicists feel it is not necessary that every facet of a theory be amenable to experiment. One might regard virtual particles as a mental or mathematical construct that helps us visualize how a field theory operates, although they are not directly detectable. What is important is that they predict phenomena that *are* directly detectable and confirm the theories.

On the other hand, if a theory predicts nothing that is directly detectable, then it has only mathematical consistency as an argument in its favor. As theories and models of the very early universe become ever farther removed from the realm of experience, some physicists argue that the traditional criterion for acceptance of a theory—that it be falsifiable, or capable of being proven wrong—is no longer tenable. Rather, we should be willing accept a theory on the basis of "meta-criteria," such as the probability that it is correct (if such a probability means anything) or even its artistic merits. To be sure, mathematical beauty has long been a driving force behind theory creation and acceptance, but proposals based on this elusive quality have turned out to be wrong as often as right.

So dramatically has the style and sociology of theoretical physics shifted in recent decades that the question inevitably arises: Have cosmologists taken to counting angels on pins? One also inevitably recollects the Yiddish proverb, "Man thinks and God laughs."

*Have we entered an era of post-empirical science? Is post-empirical science an oxymoron?*

# MULTIVERSES AND METAPHYSICS

YOU HAVE BEEN PATIENTLY holding in reserve your question about the multiverse. I have been patiently waiting. After all, no cosmology lecture would be complete without its appearance. As for an answer, there is none better than the one James Peebles, America's grand old man of cosmology, gave after a 2020 Harvard talk. Did he believe in the multiverse?

No.

End of book.

In this case, it will be. As a rule, the press and the public are fascinated by the most extreme speculations and, as a rule, on a day-to-day basis, working cosmologists are not overly concerned with them. Nevertheless, the multiverse has been in the spotlight for well over a decade, and the excitement of pondering such matters is one reason that young people become cosmologists. As mentioned in Chapter 12 and Chapter 14, the inflationary model and string theory evidently require a multiverse.

But what is such a hyper-hydra-headed universe, exactly? "Exactly" may have no place in the question, or the answer. To an extent it is a matter of semantics. If by definition "universe" means "everything," then no multiverse exists. What is typically meant by "multiverse" in modern cosmology is an ensemble of "sub-universes" with wildly differing properties. Some may be flat; most will be curved. In some, the fundamental constants of nature will be at or near the values we measure them to be. In others they will be different by orders of magnitude. In some, galaxies will exist, in others not. We live in one of them.*

---

* There is another sort of multiverse, associated with quantum mechanics. Quantum mechanics does not predict the outcome of a measurement, only the probability of a

The multiverse is the epitome of "post-empirical" science—there seems to be no way to test the multiverse concept by the traditional scientific methods sketched in the Introduction. A few proposals have been made, but none have been taken seriously enough to be actively pursued. Cosmologists search for dark matter because there is indirect observational evidence for it, but they are not searching for the multiverse, because there is no evidence for it. In his answer to the after-lecture question, Peebles reflected this position.

To be indulgent, we might ask why we are living in the particular universe we are. More specifically: Why do we observe our universe to be approximately ten billion years old?

This is the basic *anthropic* question. Robert Dicke's answer is famous: "The universe must have aged sufficiently for there to exist elements other than hydrogen, since it is well known that carbon is required to make physicists." In other

---

given outcome. Some physicists believe that at every measurement the universe splits, so that all outcomes occur, but in different universes. This is known as "the many worlds interpretation of quantum mechanics."

words, if the universe weren't at least several billion years old, we wouldn't be here to observe it. More broadly the *anthropic principle* holds that the universe as we observe it must be such as to allow life. A universe that did not produce life would not produce observers. According to the anthropic principle, the existence of life selects our particular cosmos from the multiverse.

*

When anthropic arguments became popular in the 1970s, reactions ranged from skepticism to scorn. Many physicists dismissed it as tautological; *obviously* our universe is such as to be compatible with life. An analogy offered by Dicke and Peebles, though, may make it seem less trivial. Loaded and unloaded pistols are randomly distributed to a crowd of cosmologists and they engage in a mass game of Russian roulette. Afterward, a brilliant statistician appears and discovers by exhaustive analysis that there is a high probability that any surviving cosmologist holds an unloaded pistol.

You might derisively exclaim, "Obviously!" That outcry, however, is an admission that the situation is subject to meaningful after-the-fact analysis. The main objection to the anthropic principle has always been that it cannot predict anything, and therefore fails at the fundamental requirement of a physical theory. The mass Russian roulette game makes that a little less clear; the outcome *could* have been predicted. When playing roulette with universes, admittedly, there is no way to know ahead of the game whether a given universe is loaded.

Nevertheless, a famous story in anthropic lore is that astronomer Fred Hoyle used anthropic reasoning in 1953 to predict that a certain nuclear reaction in the sun *must* exist for sufficient carbon to be produced to sustain life. Nowhere in his papers of the time, however, does he mention anthropic considerations, and the story appears to be a retrospective invention.

The situation differs with regard to American geologist Thomas Chamberlin. In the nineteenth century, a great debate raged between physicists and naturalists over the age of the earth. Darwin required untold eons to evolve the species, but physicists led by Lord Kelvin did not believe that the sun could have lasted long enough to do so, emitting energy by any known mechanism. In 1899, Chamberlin argued that Kelvin's arguments proved only that the sun was burning by some unknown source of energy locked in atoms. The Darwinians and Chamberlin turned out to be correct and the physicists wrong. Chamberlin's reasoning might have led to the discovery of nuclear reactions in the sun.

*

In recent decades, anthropic arguments have been enlisted to explain numerous features of our universe, although only in hindsight. Most relevant to our purposes are the arguments to constrain the sizes of the fingerprints of God and the cosmological constant. We have seen that the size of the fluctuations in the microwave background are roughly one part in $10^5$. If they were much larger, the matter in the universe would have collapsed into black holes. If they were much smaller, the matter would not have coalesced into galaxies and stars. In neither case would observers have arisen in such a universe.

By the same token, because it accelerates the expansion of the universe, the cosmological constant impedes matter from coalescing into galaxies. If the constant were larger than the matter content of the universe during the epoch of galaxy formation, when the observable universe was roughly a fifth its current size, no galaxies could form. The density of matter back then was about 125 times larger than at present, and so presumably the cosmological constant could not have been more than one or two orders of magnitude larger than it is today.

A major objection to anthropic arguments has always been that they rarely yield an answer with a give-or-take range less than an order of magnitude. True. On the other hand, limiting the cosmological constant to a factor of ten or so above its present value is a significant improvement over the 120 orders of magnitude mentioned in Chapter 8 based on quantum mechanical calculations.

Many physicists, even those who propose them, regard anthropic arguments as acts of desperation. They may be inevitable in an age when our quantitative theories have become so speculative; it is wrong to think that a theory filled with complicated equations necessarily means anything. One should also bear in mind that the anthropic principle is a *principle,* not a law of nature. Many principles have been enlisted throughout the history of physics to guide our thinking toward successful theories; some have proven more useful than others. The cosmological principle proved to be very successful, even if it was obviously not entirely true. But how does one test the principle of beauty? The idea of beauty in physics is often encapsulated in the idea of mathematical symmetry—that systems have regular patterns—and while the implementation of symmetry concepts has proven highly successful in particle physics, it may have outlived its usefulness. As mentioned in Chapter 14, the Large Hadron Collider has found no evidence for supersymmetry.

A famous *principle of least action* is universally accepted among physicists. The principle of least action springs from the simple idea that the shortest distance between any two points is a straight line and that light, for example, tends to travel along those straight lines. The principle says that one can obtain the equations for a given theory by minimizing a quantity known as the *action*, which is related to a system's energy. Historically, the action principle revolutionized physics and has become the route by which *all* modern theories are created. Rather than infer the correct equations from experience, one postulates an action and minimizes it to generate the theory's equations. Einstein did not consider his general relativity theory complete until he could derive the field equations from an action.* Theories of quantum gravity also begin by postulating an action. It is known, however, that sometimes the principle of least action gives the wrong answer. If we are merely postulating an action for a completely new theory, how do we know we have produced the correct equations? Especially when we cannot experimentally test the results?

*

On a spectrum ranging from the principle of beauty to the principle of least action, the anthropic principle perhaps lies closer to beauty. Moreover, I have been discussing what is known as the *weak* version of the anthropic principle, which if not tautological, does not seem unreasonable. As in Dicke's original question, it merely asks why some aspect of the universe—its age— is observed to be what it is. It assumes that the known laws of nature are what they are. Stronger

* Mathematician David Hilbert beat Einstein in this race by five days.

versions of the anthropic principle declare that the laws of nature *must* be as they are. In particular, that the fundamental constants of nature, such as $G$ and $h$, must be what we measure them to be; otherwise the universe as we know it could not exist. For example, if the constants were much different than they actually are, stars would not form, and so presumably life would not exist, either.

Physicists have a harder time accepting the strong anthropic principle because it holds echoes of the argument from design—the great clockwork of the universe must imply the existence of a watchmaker. To be sure, the strongest version of the anthropic principle, the *participatory* anthropic principle, requires the universe to eventually produce life. Physicists generally reject such ideas because they smack of teleology—the belief that things happen because of the final purpose they serve. Science has moved in the opposite direction from teleological arguments since Aristotle.

＊

Without enlisting the anthropic principle, it is presently unclear how to select viable universes from the multiverse or from the string theory landscape. The current state of affairs is undoubtedly due to the lack of experiments or observations constraining the imagination of theorists. Even if we are lucky enough to become an advanced civilization, it will remain a stretch to create universes in the laboratory to test the multiverse and the anthropic principle.

It is likely that we will never completely understand what took place at the Planck time, or before the big bang, unless the newer bounce cosmologies allow us to peer into that epoch. If our theories do not in the end provide a smooth transition to what is generally observable, we may indeed be forced to rely on mathematical consistency, and vague notions of probability and beauty to constrain them.

By the same token, it is unlikely that physicists will ever create a theory of everything. The phrase should not be taken too seriously. Even those attempting to create one would not claim it could explain why people fall in love. However, even in its more limited goal, to unite the four forces of nature, it is hardly clear how useful it would be. Along the path to a theory of everything many insights have been achieved, but a large number of scientists regard the endeavor as misguided in principle.

The most successful theories are those with limited domains of applicability. No knowledge of what took place in the earliest instants of the universe is needed to calculate the orbits of the planets. Perhaps the greatest achievement of science is that it is possible to say something without saying everything. And there is no question that a theory of everything would remain incomplete. A ten-dimensional string theory, even if accepted beyond a shadow of a doubt, would leave unanswered the question of why there are ten dimensions. No theory specifies everything about itself. Whether it is the very constants of nature, or assumptions of how the universe began, something always remains to be put in by hand. Most cosmologists would concede that they do not study cosmology to solve the ultimate mysteries of nature but to get close to them. Rest easy, then, and do not worry: future generations of cosmologists will continue to wonder . . .

**Why is there something rather than nothing?**

# FURTHER READING

Information in this book is mainly derived from technical papers and seminars that would not serve general readers. The books and articles listed below, all by respected physicists, are written in layperson's terms, although perhaps at a somewhat higher level than this work.

1. *On the experimental basis for general relativity, a new update:*

   Clifford Will and Nicolás Yunes, *Is Einstein Still Right?* (Oxford University Press, 2020).

2. *On big bang nucleosynthesis, the classic popular book:*

   Steven Weinberg, *The First Three Minutes: A Modern View of the Origin of the Universe* (Basic Books, 1977).

3. *On modern cosmology, a book covering some of the same speculative topics:*

   Martin Rees, *Before the Beginning: Our Universe and Others* (Helix Books, 1997).

4. *On observations of the CMBR, a recent book:*

   Lyman Page, *The Little Book on Cosmology* (Princeton University Press, 2020).

5. *On inflation, from the horse's mouth:*

   Alan H. Guth, *The Inflationary Universe: The Quest for a New Theory of Cosmic Origins* (Basic Books, 1999).

6. *On inflation (and strings), Roger Penrose's objections:*

   Roger Penrose, *Fashion, Faith and Fantasy in the New Physics of the Universe* (Princeton University Press, 2016).

7. *On strings, an accessible introduction:*

   Steven S. Gubser, *The Little Book of String Theory* (Princeton University Press, 2010).

8. *On research into quantum gravity (a personal view):*

   Lee Smolin, *Three Roads to Quantum Gravity* (Basic Books, 2001).

9. *On the anthropic principle (practically all anyone might want to know on the topic):*

   John D. Barrow and Frank J. Tipler, *The Anthropic Cosmological Principle* (Oxford University Press, 1986).

10. *On the inflation debate and bouncing cosmologies:*

Anna Ijjas, Paul Steinhardt, and Abraham Loeb, "Pop Goes the Universe," *Scientific American,* January 2017.

Paul Steinhardt, "The Inflation Debate," *Scientific American,* April 2011.

11. *On dark matter searches:*

Joshua Sokol, "Elena Aprile's Drive to Find Dark Matter," *Quanta,* December 20, 2016.

Daniel Bauer, "Searching for Dark Matter," *American Scientist,* September-October 2018.

12. *On gravitational waves and Mach's Principle:*

Tony Rothman, "The Secret History of Gravitational Waves," *American Scientist,* March-April 2018.

Tony Rothman, "The Forgotten Mystery of Inertia," *American Scientist,* November-December 2017.

## ACKNOWLEDGMENTS

My deep thanks to Stephen Boughn and Patti Wieser for critical readings of the book. Thanks also to the anonymous reviewers for their helpful suggestions. Of course, any remaining mistakes or oversights are my own responsibility.

# INDEX